Graphing Calculator Guidebook for the TI-82/85

PRECALCULUS
CONCEPTS IN CONTEXT

Judith Flagg Moran
Trinity College

Marsha Davis
Eastern Connecticut State University

Mary Murphy
Smith College

PWS PUBLISHING COMPANY
I(T)P   An International Thomson Publishing Company

Boston • Albany • Bonn • Cincinnati • Detroit • London • Madrid • Melbourne • Mexico City
New York • Paris • San Francisco • Singapore • Tokyo • Toronto • Washington

PWS PUBLISHING COMPANY
20 Park Plaza, Boston, MA 02116-4324

Copyright © 1996 by PWS Publishing Company,
 a division of International Thomson Publishing Inc.

All rights reserved. No part of this book may be reproduced, stored in a retrieval system, or transcribed, in any form or by any means–electronic, mechanical, photocopying, recording, or otherwise–without the prior written permission of the publisher, PWS Publishing Company.

I(T)P™
International Thomson Publishing
The trademark ITP is used under license.

ISBN: 0-534-19801-5

Printed and bound in the United States by Malloy.
95 96 97 98 99 — 10 9 8 7 6 5 4 3 2 1

Graphing Calculator Guidebook for the TI-82/85
To Accompany *Precalculus: Concepts in Context*

Table of Contents

	TI-82	TI-85
Basic Tutorial	1	64
Topics		
1. *On, Off, and Contrast*	1	64
2. *Basic Calculation and Editing*	2	66
3. *Correcting an Error*	3	67
4. *Resetting the Memory*	4	68
5. *Changing Mode Settings*	5	69
6. *Graphing*	6	73
7. *Erasing Stored Functions*	9	73
8. *Exiting the Graphing Window or a Menu*	10	73
Chapter 1	11	75
General Information		
Clearing Stored Functions	11	75
Making a Table of Values from a Formula	11	76
Lab 1: Fahrenheit		
Plotting Points	14	77
Project 1.1: The Graphing Game		
Adjusting the Viewing Window for Square Scaling	17	80
Chapter 2	19	82
General Information		
Graphing a Quadratic Function	19	82
Calculating Outputs of a Function	20	83
Programming the Quadratic Formula (Optional)	21	83
Laboratory 2A: Galileo		
Approximating the Coordinates of a Point on a Graph	26	90
Lab 2B: Electric Power		
Finding the Points of Intersection	27	92
Graphing Piecewise-Defined Functions	29	94
Project 2.5: The Least Squares Line		
Computing the Least Squares Line	30	95

Chapter 3   33   99

Lab 3: Graph-Trek
- *Functions Involving Square Roots*   33   99
- *Trigonometric Functions and the Trig Viewing Window*   33   99
- *Graphing a Family of Functions* (31)   34   100
- *The Difference Between $(-x)^2$ and $-x^2$*   35   101

Projects 3.1 and 3.2: Vertical and Horizontal Stretching and Compression
- *Entering the Family of Functions, $c \cdot f(x)$*   35   102
- *Entering the Family of Functions, $f(c \cdot x)$*   36   102

Project 3.6: Absolute Value in Functions
- *Functions Involving Absolute Value*   36   102

Chapter 4   37   103

General Information
- *Determining the Roots of a Function*   37   103

Lab 4A: Packages
- *Finding Local Maxima or Minima of Functions*   38   104

Projects 4.1 and 4.2: Exploring Polynomial Graphs
- *Determining a Good Viewing Window for a Polynomial Function*   40   105

Lab 4B: Doormats
- *Graphing a Rational Function*   41   106
- *Zooming Out* (38)

Project 4.4: Vertical Asymptotes and "Black Holes"
- *Vertical Asymptotes*   44   109
- *Holes*   45   110

Project 4.5: Long-term Behavior of Rational Functions
- *When the Degree of the Denominator Is Greater Than or Equal to the Degree of the Numerator*   46   111
- *When the Degree of the Denominator Is Less Than the the Degree of the Numerator*   46   111

Chapter 5   48   113

General Information
- *Graphing Exponential Functions*   48   113
- *Graphing Logarithmic Functions*   49   115

Lab 5A: AIDS
- *Graphing an Exponential Function Expressed in Base-e Form*   50   116

Lab 5B: Radioactive Decay
- *Graphing a Function Representing Exponential Decay*   51   117

Lab 5C: Earthquakes
- *Tracing Beyond the Viewing Window*   51   117

Chapter 6   53   119
    General Information
        Graphing Trigonometric Functions   53   119
        Problems Inherent in the Technology   53   119
    Lab 6: Daylight and SAD
        Modifying Basic Trigonometric Functions   55   121
    Project 6.4: Don't *Lean* On Me
        Away Dull Trig Tables!   56   122

Chapter 7   57   123
    General Information
        Graphing Parametric Equations   57   123
    Lab 7A: Bordeaux
        Graphing a Function of More Than One Input Variable   59   125
    Lab 7B: Bézier Curves
        Combining Two Sets of Parametric Equations   60   126
    Project 7.3: What Goes Around Comes Around
        Using Square Scaling in Parametric Mode   62   128

Preface

Graphs play an important role in providing understanding of the precalculus concepts contained in *Precalculus: Concepts in Context*. With the aid of a graphing calculator, students can graph functions, even fairly complex functions, quickly and easily. All of the labs, most of the projects, and many of the exercises in this textbook require the use of graphing technology. The *Guidebook* provides keystroke-level calculator commands and instructions that are useful in working through the labs, projects, and exercises of *Precalculus: Concepts in Context*.

The *Guidebook* is two books in one: the first provides instructions for the TI-82 graphing calculator, and the second, for the TI-85. Each guide begins with a basic tutorial followed by seven chapters corresponding to chapters in *Precalculus: Concepts in Context*. In all but one case, the same examples are used for both calculator guides. The one exception occurs in *Making a Table of Values from a Formula*. The TI-85 does not have the TI-82's Table feature that is used in this section. However, Texas Instruments does provide (upon request) a program for the TI-85 that emulates the Table feature of the TI-82.

After a basic tutorial, the *Guidebook* introduces new calculator features chapter by chapter as they are needed. We suggest that students work through the basic tutorial in class. Students should work through the sections in the *Guidebook* pertaining to the labs as part of their lab preparation. We do not assume that students or instructors will work through the entire *Guidebook* chronologically. Therefore, we've structured the *Guidebook* so that after completing the basic tutorial, you can jump around in the *Guidebook* from topic to topic.

The examples in the *Guidebook* introduce students to a wide variety of the features of their TI-82/85 graphing calculators. With the aid of a graphing calculator, many problems encountered in precalculus have multiple methods of solution. Whenever possible, we've chosen to introduce techniques that are general or graphical. For example, the POLY command on the TI-85 can be used to determine the roots of a polynomial. (Students can discover this technique in their TI-85 manual.) Instead, we discuss finding the roots of a polynomial using the ROOT command which applies to general functions and requires graphical input. Students should not feel limited by our suggestions, but should be encouraged to explore other possibilities described in their calculator's manual.

Although the calculator reduces the amount of tedious calculation, it does not, however, relieve the students from the responsibility of thinking about mathematics. Several of the examples in the *Guidebook* are designed to encourage responsible and thoughtful use of graphing technology. With this in mind, we've included several examples that illustrate limitations of graphing technology.

We would like to acknowledge Roger Nelson of Lewis and Clark College for his careful review of the *Guidebook*. In addition, we'd like to thank the teachers at various workshops and precalculus students who have tested these instructions and given us valuable feedback.

Graphing Calculator Guidebook for the TI-82/85
To Accompany *Precalculus: Concepts in Context*

This guide provides background on the TI-82/85 graphing calculator that will be useful for *Precalculus: Concepts in Context*. It consists of a basic tutorial followed by additional instructions relevant to each chapter in the text. We do not attempt to show you everything that you can do on the TI-82/85. Feel free to investigate other capabilities of your calculator by experimentation or with the aid of your TI-82/85 manual.

Turn to page 64 for instructions on the TI-85. The guide for the TI-82 begins below.

Basic Tutorial for the TI-82

As you proceed through this guide, note that specific keys that you are to press appear in a box. For example, you may be asked to press $\boxed{\text{Graph}}$. Operations corresponding to the blue or gray lettering above the keys are indicated in brackets, $\boxed{[\ \ ]}$. To access blue upper key functions, press $\boxed{\text{2nd}}$ followed by the key. To access gray upper key functions, press $\boxed{\text{ALPHA}}$ and then the key.

1. *On, Off, and Contrast*

Turn the calculator on by pressing $\boxed{\text{ON}}$.

You may need to adjust the contrast. Press $\boxed{\text{2nd}}$ followed by holding down the up arrow key $\boxed{\uparrow}$ to darken or the down arrow key $\boxed{\downarrow}$ to lighten. (The four arrow keys are located directly below the $\boxed{\text{TRACE}}$ and $\boxed{\text{GRAPH}}$ keys.)

To turn your calculator off, press $\boxed{\text{2nd}}$ $\boxed{[\text{OFF}]}$. If you forget, the calculator will automatically turn off after a period of non-use.

TI-82 Guide

2. *Basic Calculation and Editing*

The screen that displays your calculations is called the Home screen. Press $\boxed{\text{CLEAR}}$ to begin with a clear Home screen. You do <u>not</u> have to clear the screen after each computation. (If pressing $\boxed{\text{CLEAR}}$ does not clear the screen, press $\boxed{\text{2nd}}$ $\boxed{\text{[QUIT]}}$ followed by $\boxed{\text{CLEAR}}$.)

Example: Compute 3×4.
After pressing $\boxed{3}$ $\boxed{\times}$ $\boxed{4}$, press $\boxed{\text{ENTER}}$. Note that the original problem, written as 3 ∗ 4, remains on the left side of the screen and the answer appears to the right.

Example: Compute $3 + 2 \times 6$ and $(3 + 2) \times 6$.
In which order did the calculator perform the operations of addition and multiplication?

Example: Compute 8^2 and 1.05^7.
Press $\boxed{8}$ followed by $\boxed{x^2}$ $\boxed{\text{ENTER}}$. You can also compute the square of eight by pressing $\boxed{8}$ $\boxed{\wedge}$ $\boxed{2}$. Now try 1.05^7 using $\boxed{\wedge}$ $\boxed{7}$ to create the power.

Example: Compute $\sqrt{16}$.
Press $\boxed{\text{2nd}}$ $\boxed{[\sqrt{}]}$ (same key as $\boxed{x^2}$) followed by $\boxed{1}$ $\boxed{6}$ $\boxed{\text{ENTER}}$.

Example: Compute $\sqrt{-16}$.
Press $\boxed{\text{2nd}}$ $\boxed{[\sqrt{}]}$ followed by $\boxed{(\text{-})}$ (the key to the left of $\boxed{\text{ENTER}}$) $\boxed{1}$ $\boxed{6}$ $\boxed{\text{ENTER}}$.

You should have gotten the error message ERR: DOMAIN when you tried to compute $\sqrt{-16}$ because there is no real number whose square is -16. Press $\boxed{2}$ for Quit.

Example: Compute $\sqrt[5]{32}$ and $\sqrt[3]{18}$.

To compute $\sqrt[5]{32}$:

- Press $\boxed{5}$ for the fifth root.

2

Basic Tutorial

- Press [MATH] [5] for $\sqrt[x]{}$.

- Press [3] [2] [ENTER].

Next, let's approximate $\sqrt[3]{18}$ by using the $\sqrt[3]{}$ selection from the MATH menu.

- Press [MATH] [4] for $\sqrt[3]{}$.

- Press [1] [8] [ENTER].

Warning! *The TI-82 has two minus keys,* [−] *and* [(-)], *to differentiate between the operation of subtraction (such as* $3 - 2 = 1$) *and the opposite of the positive number 2, namely* -2.

Example: Compute $-2 + 5$.

To compute $-2 + 5$, press [(-)] then [2] to create the number -2. Finish the computation to get the answer 3.

3. *Correcting an Error*

We tackle three situations connected with making errors. First, we look at two examples of errors that your calculator recognizes as errors. Then we provide an example that illustrates what you can do when you discover that you have punched in an error that the calculator is able to compute.

Correcting by Deleting

Let's start by making a deliberate error: press [3] [+] [+] [2] [ENTER].
The following message will appear on your screen.

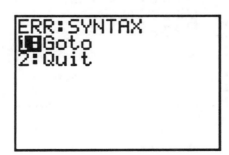

TI-82 Guide

Press [1] and the cursor will direct you to the error. Erase one of the plus signs by pressing [DEL] for delete, and then press [ENTER]. The correct answer to 3 + 2 will appear.

Correcting by Starting Over

Try entering $\sqrt{2}$ as [2] [2nd] [[√]] [ENTER].
   To correct your error, press [2] to quit. Then press [CLEAR] to clear the screen, and you can start over.

Correcting by Inserting

Finally, press [3] [+] [4] [ENTER] and suppose that you really wanted $-33 + 4$.
   Press [2nd] [[ENTRY]] to return to the previous command. Use the left arrow key [←] to position the cursor over the 3. Press [2nd] [[INS]]. (Note the cursor changes from a box to a line.) Now press [(-)] [3] [ENTER].

4. *Resetting the Memory*

Warning: Resetting the memory erases all data and programs. All calculator settings will return to the default settings. (Skip this step if you have programs or data that you do not wish to erase.)

Here's how to reset your calculator:

- Press [2nd] [[MEM]] to display the memory menu shown below.

Basic Tutorial

- Press [3] to select Reset, followed by [2] to reset the memory. The message **Mem cleared** should appear on your screen. If you can't read this message, darken the contrast. (Instructions for adjusting the contrast appear on page 1.)

5. *Changing Mode Settings*

Press [MODE]. The cursor should be blinking on NORMAL. If you have reset your calculator in Topic 4, the mode settings will match the default settings shown below.

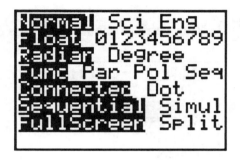

Suppose that you want all your answers to be displayed in scientific notation. Use the right arrow key [→] to highlight **Sci** (for scientific notation) and then press [ENTER]. Press [2nd] [[QUIT]] to exit the mode menu.

Example: Compute 3654 × 1781.
  The displayed answer should read: 6.507774 E 6. (See the Algebra Appendix, Section A.3(ii) for an explanation of this notation.)

To return the mode setting to NORMAL, press [MODE]; the cursor should be blinking on **NORMAL**, so press [ENTER]. (Don't exit the mode menu yet.)

Next use the down arrow key [↓] to move to the second line of the mode menu. Using the right arrow key [→], move the cursor so that it is blinking over the number 2. Press [ENTER] followed by [2nd] [[QUIT]].

Example: Compute 3.542 + 2.007.
  Did the calculator truncate or round when it displayed the answer to two-decimal place accuracy?

Press [MODE], position the cursor over **FLOAT**, and press [ENTER] to return to the default decimal setting. Press [2nd] [[QUIT]] to exit the mode menu.

5

TI-82 Guide

6. Graphing

Press [MODE]. If necessary, adjust settings to match the default settings shown in Topic 5, page 5. If you have not reset you calculator, erase any previously stored functions. (See Topic 7, *Erasing Stored Functions* (page 9), for instructions.)

Set the standard viewing window:

- Press [ZOOM] [6] for ZStandard. An empty viewing window will appear on your screen.

- Now press [WINDOW]. Your settings should match the ones shown below.

```
WINDOW FORMAT
 Xmin=-10
 Xmax=10
 Xscl=1
 Ymin=-10
 Ymax=10
 Yscl=1
```

Notice that the scaling on the x- and y-axes goes from -10 to 10 with tick marks one unit apart (since Xscl and Yscl = 1).

Graph $y = x$, $y = x^2$, and $y = x^3$ in the same viewing screen.

Step 1: Enter $y = x$.

- Press [Y=] (the key to the left of [WINDOW]). You should see the following display.

- Press [X, T, θ] (for x) and then [GRAPH]. (A pulsating line segment in the upper right corner tells you that the calculator is working on the problem. When the graph is completed, the line segment will disappear.)

Basic Tutorial

Step 2: Enter the two remaining functions.

- Press [Y=] [↓] to position the cursor opposite Y2, and then press [X, T, θ] [x^2].

- Press [↓] to position the cursor opposite Y3, and then press [X, T, θ] [^] [3].

- Now press [GRAPH].

The picture on your screen should be similar to the one shown below.

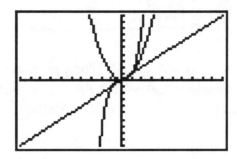

Example: Let's use the TI-82 to find where the linear function, $y = x$, and the cubic function, $y = x^3$, intersect.

Plan of action: Notice that the three points of intersection are difficult to see in the viewing screen shown above. By turning off the function $y = x^2$ and narrowing the viewing window, you'll be able to see the points of intersection more clearly. The instructions for doing this are given in Steps 1 and 2.

Step 1: Remove the graph of $y = x^2$ from the viewing screen.

- Press [Y=] to display the stored functions.

- Using the arrow keys, position the blinking cursor over the equals sign next to Y2 (pressing [↓] followed by [←] will do the trick). Press [ENTER]. This removes the highlighting over Y2's equals sign and turns the function off.

- Now, press [GRAPH] and observe that the U-shaped graph representing $y = x^2$ no longer appears in the viewing window.

<u>Note:</u> *You can turn Y2 back on by pressing* [Y=]*, positioning the cursor over Y2's equals sign, and pressing* [ENTER]*.*

7

TI-82 Guide

Step 2: Magnify the view near the center of the viewing screen.

- Press ZOOM . A portion of this menu is shown below.

- Press 2 for Zoom In. (Look for a blinking pixel at the center of your viewing window. The coordinate represented by this pixel will remain the center of the viewing window after you have zoomed in.) Press ENTER .

Step 3: Estimate the points of intersection.

Plan of action: First get acquainted with the TRACE feature on the TI-82 and then use TRACE to estimate the points of intersection.

- Press TRACE . The cursor, a box with a blinking × through the diagonals, will appear on the graph of the line $y = x$.

- Press the left arrow key ← and watch the cursor move to the left along the line. The x- and y-coordinates of the cursor's location will appear at the bottom of the screen. Now press → to move the cursor along the line in the opposite direction.

- Press ↑ or ↓ to jump back and forth between the line and the curve.

- Now you are ready to approximate the points of intersection. Position the cursor over one of the points of intersection. Read off the x- and y-values corresponding to the cursor's location. Repeat this process to approximate the other two points of intersection.

How did Zoom In affect the window settings? Before zooming in, the viewing window displayed an x-axis scaled from -10 to 10, a width of 20 units. Press WINDOW . Notice that the width of the x-scale has been reduced by a factor of 4.

Basic Tutorial

Example: Graph $y = x^3$ over the x-interval from -10 to 10.

Plan of action: The aim here is to set the y-scale so that the graph remains in the viewing window over the entire x-interval from -10 to 10. Since y has value -1000 when x is -10 and value 1000 when x is 10, you'll need to set Ymin and Ymax to -1000 and 1000, respectively.

- Press [Y=] and turn off the function $y = x$: position the cursor over the equals sign and press [ENTER] to remove the highlighting.

- Press [WINDOW].

- Use the down arrow key [↓] to position the cursor opposite Xmin. Enter -10 by pressing [(-)] [1] [0].

- Press [ENTER] or [↓] to move the cursor opposite Xmax. Continue changing the settings until your screen matches the one below.

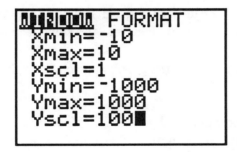

- Now press [GRAPH].

Note: The tick marks are one unit apart on the x-axis and 100 units apart on the y-axis.

7. Erasing Stored Functions

Now let's see how to erase functions that have been stored in your calculator's memory. After following the instructions in Topic 6, the functions $y = x$, $y = x^2$, and $y = x^3$ should be stored as Y1, Y2, and Y3.

TI-82 Guide

Erase $y = x$, $y = x^2$, and $y = x^3$ from your calculator's memory as follows.

- Press $\boxed{Y=}$. You should see a blinking cursor opposite Y1. Press $\boxed{\text{CLEAR}}$ to erase the function stored in Y1.

- Press $\boxed{\downarrow}$, to move the cursor opposite Y2, and then press $\boxed{\text{CLEAR}}$.

- Press $\boxed{\downarrow}$ $\boxed{\text{CLEAR}}$ to erase the function in Y3.

8. *Exiting the Graphing Window or a Menu*

Generally, pressing $\boxed{\text{2nd}}$ $\boxed{\text{[QUIT]}}$ is all that is needed to return to the Home screen from a menu or viewing window.

Example: Graph the function $y = 2x$. Then, return to the Home screen.

- Graph the function $y = 2x$ in the standard viewing window. (If your graph sits on the x-axis, you have forgotten to change to the standard viewing window.)

- Press $\boxed{\text{2nd}}$ $\boxed{\text{[QUIT]}}$ to leave the graph and return to the Home screen.

That's it! You have completed the tutorial. Now practice and experiment on your own with the calculator until you begin to feel comfortable with these basic operations. The remainder of this guide will introduce new techniques as they are needed, chapter by chapter, for your work in *Precalculus: Concepts in Context*.

Chapter 1

General Information

Clearing Stored Functions

Before you get started on a problem, you probably will want to erase any functions that are stored in your calculator's memory. The tutorial demonstrated one method for clearing functions. Here is another.

- Press [2nd] [MEM] [2] [5]. A listing of all stored functions will appear. For example, the screen below indicates that three functions, Y1, Y2, and Y3, are stored in memory.

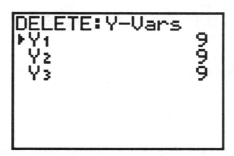

- Press [ENTER] repeatedly (once for each function stored) until all functions have been deleted.

Making a Table of Values from a Formula

The TI-82's table feature provides numeric information about a function. In the next two examples, we'll start with a column of values for the independent variable, x, and use the table feature to calculate the corresponding values for the dependent variable, y.

Example: Use your TI-82 to complete the table below for the function $y = 2x - 30$.

x	y
2.0	
6.0	
10.0	
14.0	
18.0	
22.0	
26.0	

Plan of action: First, we'll enter the function into the calculator. Then we'll set up the values for the independent variable x. Notice that the minimum x-value in the table is 2.0 and that consecutive x-values are separated by increments of 4.0. Using the TABLE command, we'll be able to generate the values for the x- and y-columns.

- Press $\boxed{\text{Y=}}$. Clear any previously stored functions and then enter the function $y = 2x - 30$ as Y1.

- Press $\boxed{\text{2nd}}$ $\boxed{\text{[TblSet]}}$ (same key as $\boxed{\text{WINDOW}}$) to access the TABLE SETUP menu.

- Set TblMin, the minimum value for x, to 2 by pressing $\boxed{2}$ $\boxed{\text{ENTER}}$.

- Set ΔTbl, the increment in x, to 4 by pressing $\boxed{4}$ $\boxed{\text{ENTER}}$.

- Check that **Auto** is highlighted for both the independent (Indpnt) and dependent (Depend) variables. (When Auto is selected, the values in the table will be created automatically.) If Auto is not highlighted, move the cursor over **Auto** and press $\boxed{\text{ENTER}}$. When you are finished, your screen should look like this:

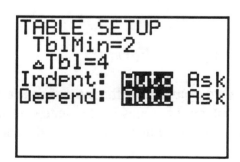

- To view the table, press $\boxed{\text{2nd}}$ $\boxed{\text{[TABLE]}}$ (same key as $\boxed{\text{GRAPH}}$). Your table should match the one that follows.

Chapter 1

```
   X    | Y1
--------|-----
   2    | -26
   6    | -18
  10    | -10
  14    | -2
  18    |  6
  22    | 14
  26    | 22
X=2
```

Notice that the dependent variable switches sign from negative to positive at some x-value between 14 and 18. In the next example, we'll pin-point the x-value corresponding to the switch in sign for the y-values.

Example: For $y = 2x - 30$, determine the x-value that corresponds to a zero y-value.

To solve this problem, we'll refine the previous table so that it includes x-values between 14.0 and 18.0 separated by increments of 1.0.

- With the table above displayed on your screen, press the down arrow key $\boxed{\downarrow}$ to move the cursor to the bottom of the table. Continue to press $\boxed{\downarrow}$ until the row containing 14 and -2 advances to the first line in the table.

- Press $\boxed{\text{2nd}}$ $\boxed{\text{[TblSet]}}$. Notice that TblMin has changed to 14 to reflect the first line of the last table displayed.

- Position the cursor to the left of **ΔTbl** = and press $\boxed{1}$ to change the value of the x-increment from 4 to 1. Then press $\boxed{\text{2nd}}$ $\boxed{\text{[TABLE]}}$ to display the refined table.

Examine the entries in the refined table. What x-value corresponds to a y-value of zero? (Did you get an x-value of 15?)

Example: Complete the table below for the function $y = \frac{1}{x}$ according to the instructions below.

x	y
4.0	
2.0	
1.0	
0.5	

13

Notice that in this table, consecutive x-values are not separated by equal increments nor are they arranged in ascending order as in the previous two examples. Here's how to generate the values for the table:

- Press $\boxed{Y=}$. Then delete any previously stored functions: position the cursor on the same line as the function and press $\boxed{\text{CLEAR}}$. Now, enter the function $y = \frac{1}{x}$ as Y1: press $\boxed{1}$ $\boxed{\div}$ $\boxed{X,T,\theta}$.

- Press $\boxed{\text{2nd}}$ $\boxed{\text{[TblSet]}}$. In order to enter the x-values manually, press the down arrow key $\boxed{\downarrow}$ to move the cursor to the line beginning Indpnt (short for independent variable). Press the right arrow key $\boxed{\rightarrow}$ to position the blinking cursor over **Ask** and then press $\boxed{\text{ENTER}}$.

- Press $\boxed{\text{2nd}}$ $\boxed{\text{[TABLE]}}$ to set up an empty table.

- Enter the first value for x: press $\boxed{4}$ $\boxed{\text{ENTER}}$. The corresponding y-value will appear automatically. Enter the remaining values for x: press $\boxed{2}$ $\boxed{\text{ENTER}}$ $\boxed{1}$ $\boxed{\text{ENTER}}$ $\boxed{.}$ $\boxed{5}$ $\boxed{\text{ENTER}}$. Your TI-82 should display the table below.

Lab 1: Fahrenheit

Plotting Points

You can use your calculator to plot the Fahrenheit-Celsius data given in the preparation for Lab 1. Then graph your guess for the formula that relates degrees Fahrenheit to degrees Celsius. This will allow you to check how closely the function specified by your formula follows the pattern of the data.

Chapter 1

Example: Plot the data in the following table and then overlay the graph of $y = 18x + 85$.

Sample Data	
x	y
-2	40
-1	60
1	100
3	140

Step 1: Adjust the viewing window as follows.

- Press $\boxed{\text{WINDOW}}$.

- Make sure that you choose a value for Xmin that is smaller than the x-coordinates in the table and a value for Xmax that is larger than the x-coordinates.

- Similarly, select appropriate settings for Ymin and Ymax.

- Decide on the spacing of the tick marks and set Xscl and Yscl accordingly. (What would be the disadvantage of setting Yscl=1? How many tick marks would appear between 40 and 140?)

Step 2: Clear or turn off any stored functions. (See *Clearing Stored Function* on page 11.)

Plan of action: The TI-82 has six lists for storing data, L1 – L6. We'll store data from the x-column in L1 and the y-column in L2. However, before entering the data in these lists, we'll first erase any previously stored data. The instructions for erasing and entering the data are outlined in Steps 3 and 4, respectively.

Step 3: Check to see if any data has been stored in lists L1 and L2. If it has, erase the data as follows.

- Press $\boxed{\text{2nd}}$ $\boxed{\text{[MEM]}}$ $\boxed{2}$ for Delete. Then press $\boxed{3}$ $\boxed{\text{ENTER}}$ to see which lists contain stored data. For example, if data is stored in L1 and L2, your screen would match the one below.

15

TI-82 Guide

- If data has been stored in L1 and L2, press ENTER twice to clear the data. Otherwise, skip to Step 4.

Step 4: Enter the sample data.

- Press STAT 1 for Edit. Your screen should look similar to the one below.

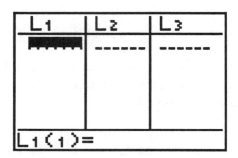

The screen above shows three empty lists, L1 – L3. At the bottom of the screen, **L1(1) =** indicates that the first entry in L1 is highlighted. Press the right arrow key → to highlight the first entry in L2 and observe that the bottom of the screen now displays **L2(1) =**. Press the left arrow key ← to return to L1(1).

- Enter the x-data in L1: press (-) 2 ENTER (-) 1 ENTER 1 ENTER 3 ENTER .

- Press the right arrow key → to move the cursor to the top of list L2. Enter the data from the y-column in L2.

Step 5: Plot the sample data.

- Press 2nd [STAT PLOT] (same key as Y=). The menu that appears gives you a choice of three plots.

Chapter 1

- Press 1 for Plot 1. Your screen should be similar to the one below.

- Press ENTER to highlight **On**. Adjust the remaining settings to match the ones above by positioning the cursor over the setting and pressing ENTER.

- Plot the sample data points by pressing GRAPH.

Step 6: Add the graph of $y = 18x + 85$ to the plot of the sample data.

- Press Y= and erase any previously stored functions. (See Topic 7, page 9 for help in clearing stored functions.) Enter the function $y = 18x + 85$ as Y1.

- Finally, press GRAPH. (Your display should show four data points that lie fairly close to the line.)

Step 7: Turn off all Plots. (If you skip this step, the sample data points will be superimposed on your next graph.)

- Press 2nd [STAT PLOT] 4 for PlotsOff.

- Press ENTER. Your calculator should respond with the message **Done**.

Project 1.1: The Graphing Game

Adjusting the Viewing Window for Square Scaling

The viewing screen on your calculator is a rectangle. Therefore, if you use the standard window, the tick marks on the y-axis will be closer together than those on the x-axis. For square scaling, we want the distance between 0 and 1 on the x-axis to be the same as the distance between 0 and 1 on the y-axis.

TI-82 Guide

First, we'll observe the graph of the line $y = x$ in the standard window and then we'll switch to square scaling.

Step 1: Graph $y = x$ in the standard viewing window.

- Erase any previously stored functions. (See instructions on page 11.)

- Press [Y=] [X,T,θ].

- Press [ZOOM] [6] for ZStandard.

Observe the spacing between the tick marks on the x- and y-axes. Notice that the tick marks on the y-axis are closer together than the tick marks on the x-axis.

Step 2: Change to a square viewing window.

- Press [ZOOM] [5] to select ZSquare. When the graph appears on your screen, observe the equal distance between tick marks on the two axes.

- Press [WINDOW] and note the changes for Xmin and Xmax from the standard window settings.

- Press [GRAPH] to return to the graph of $y = x$.

In the square viewing window, the line should appear to be inclined at a 45° angle, cutting the 90° angle made by the intersection of the x- and y-axes in half. The distances between the tick marks on the x- and y-axes should be the same.

Chapter 2

General Information

Graphing a Quadratic Function

When you graph a quadratic function, it is important to experiment with various viewing windows to ensure that you have captured all the important features of the graph on your screen.

Example: Graph $y = 2x^2 - 2x + 12$ using the three different viewing windows specified in the directions below.

Start by entering the function.

- Clear all previously entered functions from your calculator's memory. (If you need help, instructions are on page 11.)

- Press $\boxed{Y=}$ and enter $2x^2 - 2x + 12$. (Remember to use the blue subtraction key when entering $-2x$.)

- Now press $\boxed{\text{WINDOW}}$.

Window 1: Adjust the settings to match those shown below and then press $\boxed{\text{GRAPH}}$.

```
WINDOW FORMAT
Xmin=-2
Xmax=3
Xscl=1
Ymin=10
Ymax=20
Yscl=1
```

Your graph should be the familiar ∪-shape of a parabola.

Next, we view the graph of this quadratic function in two other windows. In each case, the viewing window selected fails to show key features of the parabola.

TI-82 Guide

Window 2: Change the Xmin setting to 2 as follows.

- Press WINDOW.

- Press the down arrow key ↓ to move the cursor opposite Xmin and change the value for Xmin to 2.

- Now press GRAPH.

Notice that in Window 2 the graph of $y = 2x^2 - 2x + 12$ looks more like a line than a parabola.

Window 3: Change to the standard viewing window by pressing ZOOM 6.

What do you see? In general, if you ask the calculator to graph a function, and just see empty axes, you probably are looking at a part of the plane that contains none of the graph.

Calculating Outputs of a Function

Example: Find the output of $f(x) = x^2 - x + 6$ when x has value 4, and then locate the point $(4, f(4))$ on the graph of $y = f(x)$.

- Press Y= and erase any previously stored functions.

- Enter $x^2 - x + 6$ opposite Y1.

- Press WINDOW and then ↓. Adjust the window settings to match the ones below. Then press GRAPH.

```
WINDOW FORMAT
 Xmin=-10
 Xmax=10
 Xscl=1
 Ymin=0
 Ymax=20
 Yscl=1
```

- Press 2nd [CALC] (same key as TRACE) 1 for value.

Chapter 2

- Enter the value for x: press $\boxed{4}$ $\boxed{\text{ENTER}}$. The corresponding y-value, $f(4)$, will appear at the bottom of your screen. In addition, the cursor (a plus sign with a blinking pixel at the center) will locate the point $(4, f(4))$ on the graph.

Programming the Quadratic Formula (Optional)

In order to solve inequalities and equations algebraically in Lab 2B and in some of the Chapter 2 exercises, you will need to use the quadratic formula. The steps involved in applying the quadratic formula can be stored as a program in your calculator.

The program QUADFORM, designed to calculate the discriminant and solutions to a quadratic equation of the form

$$Ax^2 + Bx + C = 0,$$

is given in the table. Instructions on entering and running this program follow. (We have placed a line number to the right of each command. Do <u>not</u> enter these numbers as part of your program.)

Quadratic Formula Program

Command	Line
NAME= QUADFORM	0
:Disp "INPUT A,B,C"	1
:Input A	2
:Input B	3
:Input C	4
:$B^2 - 4AC \to D$	5
:If $D < 0$	6
:Goto 1	7
:$(-B - \sqrt{D})/(2A) \to E$	8
:Disp "X="	9
:Disp E	10
:$(-B + \sqrt{D})/(2A) \to F$	11
:Disp "X="	12
:Disp F	13
:Lbl 1	14
:Disp "D="	15
:Disp D	16

TI-82 Guide

Step 1: Preliminary Advice.

- If you wind up in a menu that you had not intended to be in, press [2nd] [QUIT]. To return to the program you are writing, press [PRGM], use [→] to highlight **EDIT**, and then enter your program's number.

- If you need to insert a blank line below another line, move the cursor to the end of the line and press [2nd] [[INS]] [ENTER]. You can delete a line with [DEL].

- Don't be afraid to turn to the section on programming in your TI-82 manual for additional help.

Warning: *Do not reset your calculator after entering your program or you will lose your program.*

Step 2: Entering the Program.

After you have entered a command line, press [ENTER] to move to the next command line. The TI-82 will automatically insert a colon at the start of the new command line. Instructions relevant to individual command lines in QUADFORM follow.

To begin, press [PRGM], use [→] to highlight **NEW**, and then press [1].

Line 0: On your screen you should see **Name=** followed by a blinking cursor with an A inside. (This style cursor indicates that the [ALPHA] key has been activated.) Type in the name of the program QUADFORM. (The letters appear in alphabetical order on your calculator with [[A]] the gray upper function of the [MATH] key, and [[Q]] the upper function of the [9] key.) Remember that you must press [ENTER] to move to the next command line.

Lines 1 − 4. When you run this program, you will need to input the values of the coefficients of the quadratic function, $Ax^2 + Bx + C$. Create the input request prompts as follows:

- Press [PRGM], use [→] to highlight **I/O**, and then press [3] for **Disp**(lay).

- To enter the remainder of command line 1:

  Press [2nd] [[ALPHA]] to lock on the [ALPHA] key. Press [["]] [[I]] [[N]] [[P]] [[U]] [[T]] [[␣]] (same key as [0]) [[A]].

Chapter 2

Press [ALPHA] to release the Alpha-lock so that you can enter the comma.

Press [,] [ALPHA] [B] [,] [ALPHA] [C] [ALPHA] ["].

Remember to press [ENTER] to move to the next command line.

- To create lines 2 − 4, you will need to access the **Input** command. Press [PRGM], use [→] to highlight **I/O**, and then press [1].

Line 5. The value of the discriminant is computed. To enter the formula for the discriminant:

- Press [ALPHA] [B] x^2 [−] [4] [ALPHA] [A] [C]

- Then press [STO▷] for store as, [ALPHA] [D].

Here's what lines 6 and 7 do: If the discriminant is negative, there are no real solutions to the quadratic equation. In this case, no further calculations will be made and only the discriminant will be displayed. This is accomplished as follows. When D is negative, the **If** statement in line 6 is true and the program executes the **Goto** command in line 7. Executing the **Goto** command causes the program to jump down to **Lbl 1** (line 14) thus bypassing any further calculations.

Enter line 6 as follows.

- To key in **If**, press [PRGM] [1]. Then press [ALPHA] [D].

- To insert the inequality sign, <, press [2nd] [TEST] [5]. Press [0] to complete command line 6.

Now enter line 7:

- To key in **Goto**, press [PRGM]. The ↓ next to item 7 indicates that there are additional items in this menu. Press [↓] to see the other choices. You will find Goto listed as choice 0. Press [0].

- Complete line 7 by pressing [1]

Lines 8 − 13: If the if-statement in line 6 is false, the program jumps to line 8. This happens when the discriminant is positive or zero and there are two real solutions. (Note the two solutions are the same when D = 0.) Lines 8 − 13 compute and display the two solutions.

TI-82 Guide

- In line 8, remember to use the $\boxed{(-)}$ key for the opposite of B and the subtraction key $\boxed{-}$ between B and $\sqrt{D}$.

- In line 9, press $\boxed{\text{2nd}}$ $\boxed{\text{[TEST]}}$ $\boxed{1}$ to get the equals sign.

Lines 14 – 16. **Lbl 1** marks the place where the program is resumed if the discriminant is negative. Lines 15 and 16 display the value of the discriminant.

- To key in **Lbl**, press $\boxed{\text{PRGM}}$ $\boxed{9}$ and then press $\boxed{1}$ to complete line 14.

- After you have finished entering lines 15 and 16, press $\boxed{\text{2nd}}$ $\boxed{\text{[QUIT]}}$ to return to the Home screen.

Step 3: Debugging the Program.

This is best done by running the program and letting the TI-82 identify the errors. Then, after you get the program to run, take an equation and find the solutions by hand. Test QUADFORM using this equation so you can check that it is giving you the correct solutions.

(1) Let's try to run QUADFORM using the equation

$$2x^2 + 5x - 3 = 0.$$

If you encounter an error along the way, jump ahead to the instructions in (2).

- To run QUADFORM, press $\boxed{\text{PRGM}}$, followed by the number of your program. Your calculator will respond **prgmQUADFORM**. Press $\boxed{\text{ENTER}}$ to execute QUADFORM. You should get the screen below.

- Enter the coefficients of the quadratic: press $\boxed{2}$ $\boxed{\text{ENTER}}$ $\boxed{5}$ $\boxed{\text{ENTER}}$ $\boxed{(-)}$ $\boxed{3}$ $\boxed{\text{ENTER}}$. If all goes well, your screen should match the one that follows, showing two roots, -3 and $.5$, and a discriminant of 49.

Chapter 2

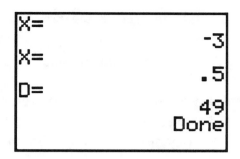

(2) Correct any errors you have made in entering your program.

- If you have made errors that the calculator recognizes as errors, you will get an error message when you execute the program. For example, if in line 9 you used the subtraction key for -B, the following error would appear when you tried to execute the program.

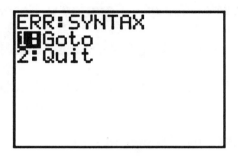

If you press [1] for Goto, you will return to the program and the cursor will mark the location of the first error that your TI-82 encountered in executing the program. Correct the error and try running the program again. Repeat this process until all errors have been corrected.

Step 4: Transferring the program from one calculator to another.

After entering a lengthy program, you may want to transfer your program to another student's calculator. You will need your calculator's link cable in order to do this.

- Connect the calculator with the program to the calculator without the program using the link cable. (There is a small hole for this cable at the bottom of your calculator.) Make sure that the cable connectors are pushed all the way in to form a good connection.

- Turn both calculators on.

25

TI-82 Guide

- The person who wants to receive the program should now press [2nd] [LINK]. Then press the right arrow key [→] to highlight **RECEIVE** and press [ENTER]. The message **Waiting** should appear on their screen.

- Next, the person who wants to send the program should press [2nd] [LINK] [3]. To select the program QUADFORM, position the black marker opposite QUADFORM and press [ENTER]. Now, use the right arrow key [→] to highlight **TRANSMIT** and press [ENTER]. When the transmission is complete, the message **Done** will appear on both the sender's and receiver's screens.

Laboratory 2A: Galileo

Approximating the Coordinates of a Point on a Graph.

Example: Approximate the vertex (turning point) of the parabola $y = 2x^2 - 2x + 12$.

Step 1: Graph the function $y = 2x^2 - 2x + 12$ and get a rough approximation for the vertex.

- Press [Y=] and enter the function $2x^2 - 2x + 12$.

- Press [WINDOW] and adjust the settings to match the ones on the screen below.

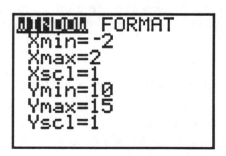

- Press [TRACE]. Use the right [→] and left [←] arrow keys to position the cursor at the vertex of the parabola. Then, read the approximate coordinates of the vertex at the bottom of your screen.

Step 2: Next, let's improve the accuracy of the estimate by magnifying the portion of the graph near the vertex of the parabola as follows.

- Press ZOOM 1 for ZBox. A free floating cursor (four dots surrounding a blinking center dot) will appear in the center of your screen. Press the left arrow key ← to move the cursor off the y-axis so that you can see it clearly.

- Draw a box surrounding the vertex of the parabola as follows. Use the arrow keys to position the cursor at the upper left-hand corner of your box. (Refer to the picture below.) Press ENTER. Now, press the right → and down ↓ arrow keys to move the lower right-hand corner of the box. When the picture on your screen resembles the one below press ENTER. (The coordinates of the lower right corner of the box appear at the bottom of the screen. The coordinates on your screen will probably differ from those below.)

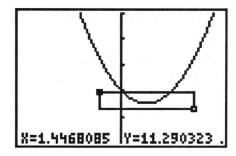

- Again, use TRACE to approximate the coordinates of the vertex of the parabola.

Notice that, in the picture above, the box drawn about the vertex of the parabola is short and wide. When we zoom in with a short-wide box, the sides of the parabola near the vertex become quite steep. If we had magnified a section of the parabola near the vertex using a square-shaped box or a tall-narrow box, the magnified section would have been flatter. It is difficult to pin-point the vertex when the section of graph surrounding the vertex is fairly flat.

Lab 2B: Electric Power

Finding the Points of Intersection

Example: Find the points where the graphs of $f(x) = -2x^2 + 2x + 12$ and $g(x) = -3x + 14$ intersect.

You could approximate the solutions to this problem using the techniques outlined in the example on page 7. However, the TI-82's intersect command will provide exact solutions when the coordinates of the point of intersection have finite decimal

TI-82 Guide

representations that are short enough to fit on the screen's display. In other cases, you will get a very good approximation without having to zoom in on a section of the graph.

Step 1: Find a viewing window that gives a clear view of the points of intersection.

- Enter the two functions into your calculator.

- Adjust the WINDOW settings so that you can see both points of intersection. (Hint: You might start with the standard viewing window, ZOOM 6, and then adjust the WINDOW settings after viewing the graph.)

Step 2: Approximate the coordinates of one of the points of intersection.

- Press 2nd [CALC] (the key above TRACE) and then press 5 for intersect. The image on your screen should be similar the one shown below.

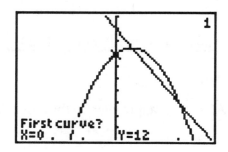

- Press the up ↑ and down ↓ arrow keys. The cursor will jump back and forth from the line to the parabola. Position the cursor on the line (We'll designate the line to be the first curve.) and press ENTER. Then position the cursor on the parabola (the second curve), and again press ENTER.

- Now you must provide your calculator with a guess for the point of intersection. Press → or ← to move the cursor to one of the points of intersection. Press ENTER. Read off the approximate coordinates for this point of intersection.

Step 3: Find the coordinates of the second point of intersection.
Repeat the process outlined by the previous three bullets to approximate the coordinates of the other point of intersection.

If you have done everything correctly, you will find that two graphs intersect at (.5, 12.5) and (2, 8).

Chapter 2

Graphing Piecewise-Defined Functions

Example: Graph the piecewise-defined function $f(x) = \begin{cases} x-4 & \text{if } x > 4 \\ -x+4 & \text{if } x \leq 4 \end{cases}$

The graph of $f(x)$ consists of two half-lines pieced together. You'll want the graph of $y = x - 4$ when x-values are greater than 4 and $y = -x + 4$ when x-values are less than or equal to 4.

Step 1: Erase any previously stored functions. Then press $\boxed{\text{ZOOM}}$ $\boxed{6}$ to set the standard viewing window.

Step 2: Enter the functions that you want to piece together.

- Press $\boxed{\text{Y=}}$.

- Enter $y = x - 4$ opposite Y1.

- Press $\boxed{\downarrow}$. Enter $y = -x + 4$ opposite Y2.

- Press $\boxed{\text{GRAPH}}$. Your graph should look like an ×.

Step 3: Form the function $f(x)$. This is where you'll piece together the graphs of Y1 and Y2.

- Press $\boxed{\text{Y=}}$ and $\boxed{\downarrow}$ to move the cursor opposite Y3.

Enter $f(x)$ as Y3 as follows:

- To insert Y1, press $\boxed{\text{2nd}}$ $\boxed{\text{[Y-VARS]}}$ $\boxed{1}$ $\boxed{1}$.

- Press $\boxed{\times}$. Next, enter the condition that governs when to use Y1: press $\boxed{(}$ $\boxed{\text{X,T,}\theta}$ $\boxed{\text{2nd}}$ $\boxed{\text{[TEST]}}$ $\boxed{3}$ $\boxed{4}$ $\boxed{)}$.

- Press $\boxed{+}$.

- To insert Y2, press $\boxed{\text{2nd}}$ $\boxed{\text{[Y-VARS]}}$ $\boxed{1}$ $\boxed{2}$.

- Press $\boxed{\times}$. Then enter the condition that governs when to use Y2: press $\boxed{(}$ $\boxed{X,T,\theta}$ $\boxed{2nd}$ $\boxed{[TEST]}$ $\boxed{6}$ $\boxed{4}$ $\boxed{)}$. When you have completed this step, your screen should match the one below.

```
Y1■X-4
Y2■-X+4
Y3■Y1*(X>4)+Y2*(
X≤4)
Y4=
Y5=
Y6=
Y7=
```

Here's how your calculator interprets the information you've just entered as a piecewise defined function. The calculator assigns the expression $(x > 4)$ the value 1 when the inequality is true (in other words, when the input variable, x, is greater than 4); when the inequality is false, the calculator sets the expression $x \leq 4$ equal to 0. Thus, for $x > 4$, the function Y3 is equivalent to:

$$Y3 = (x-4)(1) + (-x+4)(0) = x - 4.$$

And when $x \leq 4$ the function Y3 is equivalent to:

$$Y3 = (x-4)(0) + (-x+4)(1) = -x + 4.$$

Step 4: Graph $f(x)$.

- Turn off Y1 and Y2: Position the cursor over the equals sign opposite Y1. Press $\boxed{\text{ENTER}}$. Then, position the cursor over the equals sign opposite Y2 and press $\boxed{\text{ENTER}}$.

- Press $\boxed{\text{GRAPH}}$. The graph of $f(x)$ should look V-shaped.

Project 2.5: The Least Squares Line

Computing the Least Squares Line

If your data, when plotted, lies exactly on a line, you can use algebra to determine the equation of the line. However, real data seldom fall precisely on a line. Instead, the plotted data may exhibit a roughly linear pattern. The least squares line (also called the regression line) is a line that statisticians frequently use when describing a linear trend in data.

Example: Use the least squares line to describe the linear pattern in the data below.

x	y
-3.0	-6.3
-2.0	-2.8
1.2	2.0
2.0	4.1
3.1	4.7
4.2	5.6

Step 1: Enter the data from the table above into lists L1 and L2. (Refer to *Plotting Points* on pages 14 – 17.) Erase any stored functions. Then plot the data.

Step 2: Next, we'll find the equation for the least squares line. Statisticians specify the form of this line as follows: $y = a + bx$. Here b represents the slope of the line and a, the y-intercept.

- Press $\boxed{\text{STAT}}$, use $\boxed{\rightarrow}$ to highlight CALC, and then press $\boxed{9}$, for LinReg(a+bx), $\boxed{\text{ENTER}}$. The following display will appear on your viewing screen.

```
LinReg
 y=a+bx
 a=-.2829472303
 b=1.635942433
 r=.9815815587
```

- Obtain the least squares line (regression line) for the data in the table by substituting the calculated values for a and b into the equation $y = a + bx$. Write your equation in the space provided below.

$y = $ _____ + _____ x

Step 3. Graph the least squares line and the data in the same viewing window.

TI-82 Guide

- Press $\boxed{Y=}$. Then enter the function that you have written for Step 2; or press the following keys and the TI-82 will insert the regression equation for you: $\boxed{VARS}$ $\boxed{5}$ for Statistics, use the right arrow key $\boxed{\rightarrow}$ to highlight **EQ**, and then press $\boxed{7}$ for RegEQ.

- Press $\boxed{2nd}$ $\boxed{[STAT\ PLOT]}$ $\boxed{1}$. Turn Plot 1 **On** and adjust the remainder of the settings to match the menu on page 17.

- To plot both the data and least squares line, press $\boxed{ZOOM}$ $\boxed{9}$ for ZoomStat.

  <u>Note:</u> *ZoomStat adjusts the WINDOW settings so that all data points are visible on the graphing screen.*

The picture on your calculator's screen should match the one below.

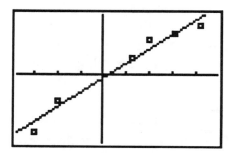

Warning: *Press* $\boxed{2nd}$ $\boxed{[STAT\ PLOT]}$ $\boxed{4}$ $\boxed{ENTER}$ *to turn off all plots. Otherwise, the data shown above will be plotted the next time you graph a function.*

Chapter 3

Lab 3: Graph-Trek

In Lab 3 you will be investigating the effect that certain algebraic modifications, such as adding a constant to the input variable, have on the graph of a function. You'll want to experiment using several different functions. We've provided some functions and algebraic modifications that you might want to consider.

Functions Involving Square Roots

Example: Compare the graphs of $y = \sqrt{x}$, $y = \sqrt{x+2}$, and $y = \sqrt{x} + 2$.

- Clear any previously stored functions. (Refer to the instructions on page 11.)

- Enter $y = \sqrt{x}$: press $\boxed{Y=}$ $\boxed{2nd}$ $\boxed{[\sqrt{}]}$ $\boxed{X,T,\theta}$.

- Enter $y = \sqrt{x+2}$: press $\boxed{Y=}$ $\boxed{2nd}$ $\boxed{[\sqrt{}]}$ $\boxed{(}$ $\boxed{X,T,\theta}$ $\boxed{+}$ $\boxed{2}$ $\boxed{)}$.

- Finally, enter the function $y = \sqrt{x} + 2$: press $\boxed{Y=}$ $\boxed{2nd}$ $\boxed{[\sqrt{}]}$ $\boxed{X,T,\theta}$ $\boxed{+}$ $\boxed{2}$. (Notice that the parentheses are the only difference between the previous function and this one.)

- Press $\boxed{GRAPH}$ to view the graphs of the three functions that you have entered. (You may need to adjust the WINDOW settings.)

Trigonometric Functions and the Trig Viewing Window

Locate the $\boxed{SIN}$, $\boxed{COS}$, and $\boxed{TAN}$ keys on your calculator. These keys will allow you to study the graphs of the sine, cosine, and tangent functions before the functions are formally introduced in Chapter 6. Prior to entering any of these functions, press $\boxed{MODE}$. If **Radian** is not highlighted, move the cursor to Radian and press $\boxed{ENTER}$.

Example: Examine the graphs of $y = \sin(x)$ and $y = \sin(x+2)$.

- Enter the function $\sin(x)$ opposite Y1: press $\boxed{SIN}$ $\boxed{X,T,\theta}$.

TI-82 Guide

- Next, enter the function $\sin(x+2)$ opposite Y2: press [SIN] [(] [X,T,θ] [+] [2] [)]

- Press [ZOOM] [7] for the trig viewing window. You should see two wavy curves. (If you don't, press [MODE]. You have probably forgotten to highlight Radian.)

- Finally, press [WINDOW] and observe the settings for the trigonometric window.

Graphing a Family of Functions

Using your calculator's list capabilities, you can substitute each value in a given list for a constant in an algebraic formula. This feature allows you to graph an entire family of functions quickly. On the TI-82, you specify a list by enclosing the members of the list in brackets: { }.

Example: Graph the family of functions $y = (x+1)^2$, $y = (x+2)^2$, and $y = (x+3)^2$.

- Clear any previously stored functions from your calculator's memory. (Refer to *Clearing Stored Functions* on page 11.) Then press [WINDOW] and change the settings to match those shown below.

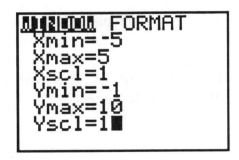

- Press [MODE]. Check that Sequential is highlighted.

- Enter the three functions by specifying the constants, 1, 2, and 3, in a list as follows. Press [Y=] [(] [X,T,θ] [+] [2nd] [[{]] [1] [,] [2] [,] [3] [2nd] [[}]] [)] [x^2]. If you have entered the family of functions correctly, your screen will match this one:

34

```
Y1=(X+{1,2,3})²
Y2=
Y3=
Y4=
Y5=
Y6=
Y7=
Y8=
```

- Press GRAPH. Watch as the three functions are graphed one after the other.

The Difference Between $(-x)^2$ and $-x^2$

Example: Graph the functions $f(x) = (-x)^2$ and $g(x) = -x^2$ in the standard viewing window.

- Clear any previously stored functions, then press ZOOM 6 to set up the standard viewing window.

- After pressing Y=, enter the function $f(x) = (-x)^2$: press ((-) X,T,θ) x^2.

- Now enter the function $f(x) = -x^2$: press (-) X,T,θ x^2.

Which of these functions has negative outputs?

Projects 3.1 and 3.2: Vertical and Horizontal Stretching and Compression

This section assumes that you are familiar with the TI-82's list capabilities. You may want to review *Graphing a Family of Functions* on page 34 before working through this section.

Entering the Family of Functions, $c \cdot f(x)$

Example: To graph the family of functions $y = x^2$, $y = 2x^2$, and $y = .5x^2$, enter the family as $\{1, 2, .5\}x^2$.

TI-82 Guide

Entering the Family of Functions, f(c · x)

Example: To graph the family of functions $y = x^2$, $y = (2x)^2$, and $y = (.5x)^2$, enter the family as $(\{1, 2, .5\}x)^2$.

Project 3.6: Absolute Value in Functions

Functions Involving Absolute Value

Example: Graph $y = |x|$ and $y = |x+2|$ in the standard viewing window.

- Clear any functions from your calculator's memory. (Refer to *Clearing Stored Functions* on page 11.) Then set up the standard viewing window by pressing $\boxed{\text{ZOOM}}\ \boxed{6}$.

- To enter $y = |x|$ as Y1: press $\boxed{\text{Y=}}\ \boxed{\text{2nd}}\ \boxed{\text{[ABS]}}\ \boxed{\text{X,T,}\theta}$.

- To enter $y = |x+2|$ as Y2: press $\boxed{\downarrow}$ to move the cursor opposite Y2, then press $\boxed{\text{2nd}}\ \boxed{\text{[ABS]}}\ \boxed{(}\ \boxed{\text{X,T,}\theta}\ \boxed{+}\ \boxed{2}\ \boxed{)}$.

- Press $\boxed{\text{GRAPH}}$ to view the graphs of the two functions.

Note: *When you want the absolute value of an expression, you must enclose the entire expression in parentheses.*

Chapter 4

General Information

Determining the Roots of a Function

Determining the roots (zeros, or x-intercepts) of a function using factoring often requires considerable skill; and most polynomials can't be factored easily, if at all. However, your calculator can take much of the drudgery out of finding roots.

Example: Find the roots of the polynomial function $g(x) = \frac{1}{2}x^4 + 2x^3 + x^2 - 3x - 5$.

Step 1: Graph $g(x)$ in the standard viewing window.

- Clear any previously stored functions. (See *Clearing Stored Functions* on page 11.)

- Press $\boxed{\text{Y=}}$ and enter $g(x)$ as Y1.

- Press $\boxed{\text{ZOOM}}$ $\boxed{6}$ to view the graph.

Step 2: Let's approximate the negative root first. (You could zoom in on this intercept and then use TRACE to estimate the x-coordinate. However, here is a way to estimate the root without zooming in.)

- Press $\boxed{\text{2nd}}$ $\boxed{\text{[CALC]}}$ $\boxed{2}$ for root.

- Select a lower bound: use $\boxed{\leftarrow}$ to move the cursor along the graph until it lies to the left of the negative root and then press $\boxed{\text{ENTER}}$.

- Select an upper bound: use $\boxed{\rightarrow}$ to move the cursor to the right of the negative root (<u>but still to the left of the positive root</u>) and then press $\boxed{\text{ENTER}}$.

- Guess the root: use $\boxed{\leftarrow}$ and/or $\boxed{\rightarrow}$ to move the cursor as close to the negative root as possible and then press $\boxed{\text{ENTER}}$. Your screen should look similar to the one shown below.

TI-82 Guide

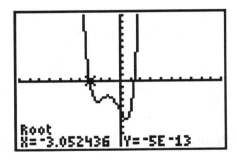

- Read off the approximate value of the root (the x-coordinate) from the bottom of your viewing screen. The corresponding y-value will be very close to zero. (Note that -5E-13 is $-.0000000000005$.)

Step 3: Next, approximate the positive root: press 2nd [CALC] 2 and then adapt the instructions in Step 2. If you have adapted the instructions correctly, you should get approximately 1.39 for this root.

Lab 4A: Packages

Finding Local Maxima or Minima of Functions

In Lab 2A, you found the vertex of a parabola using a procedure requiring a combination of ZOOM Box and TRACE. (For details, refer to the example that begins on page 26.) The coordinates of any function's turning points could be estimated using this method. One drawback to this procedure is that you must frequently apply it several times in succession before you can obtain the desired accuracy. Here is another method for finding local maxima or minima of a function.

Example: Let's estimate the local maximum and minimum of the cubic function

$$f(x) = x^3 - 4x^2 + 2x - 4.$$

Step 1: Graph $f(x)$ using a viewing window that gives you a clear view of the two turning points (one peak and one valley) of the graph.

Step 2: Approximate the coordinates of the turning point associated with the local maximum (y-coordinate of the peak on the graph) as follows.

Chapter 4

- Press [2nd] [[CALC]] (same key as [GRAPH]) and then press [4] for maximum. You're screen should look similar to the one below.

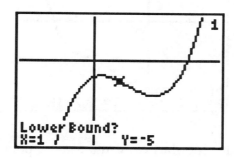

Next, you need to specify an x-interval that contains the x-coordinate associated with the local maximum of the function:

- Specify a lower bound for this interval. Use [←] to select an x-value that is less than the x-value of the turning point of the graph's peak. Then press [ENTER]. A black triangle that points to the right will mark this lower bound.

- Specify an upper bound for this interval. Use the [→] to specify an x-value that is larger than the x-value of the turning point of the graph's peak. Then press [Enter]. A black triangle that points to the left will mark this upper bound. At this point, the image on your screen should be similar to the one below.

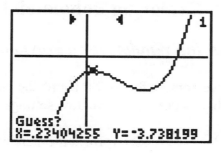

- Provide a guess for the x-value corresponding to the local maximum. Use the right or left arrow keys to position the cursor at the top of the turning point and then press [ENTER]. After a brief wait the coordinates of this turning point will appear at the bottom of your screen. Compare your answer with the one shown on the screen that follows.

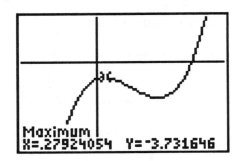

Step 3: Approximate the coordinates of the turning point associated with the local minimum (y-coordinate of the valley on the graph).

- Press [2nd] [CALC] (same key as [GRAPH]) and then press [3] for minimum.

- To determine the value of the local minimum, follow the last three instructions of Step 2: specify a lower bound and an upper bound containing the turning point (the valley), provide a guess for this turning point, and then read off the coordinates at the bottom of your screen. You should get a value for y that is close to -8.42.

Projects 4.1 and 4.2: Exploring Polynomial Graphs

Determining a Good Viewing Window for a Polynomial Function

It is frequently helpful to have some information about the range of a function's outputs prior to determining reasonable values for the window settings.

Example: Graph $f(x) = 5x^3 - 9x^2 - 40x + 15$ in the standard viewing window.

After viewing the graph of $f(x)$ in the standard viewing window, you should realize that the settings for Ymin and Ymax need to be changed. But, by how much should you decrease Ymin or increase Ymax in order to get a more comprehensive view of this graph? One help is to evaluate the function at a couple of specific values.

- Approximate the outputs $f(-1)$ and $f(2)$. (See *Calculating Outputs of a Function* on page 20.) Did you get 41 for $f(-1)$?

- Adjust the settings on the viewing window so that yMin is somewhat smaller than $f(2)$ and yMax is somewhat larger than $f(-1)$.

Chapter 4

The graph of $f(x)$ should now look much more like a classic cubic with two turning points.

Lab 4B: Doormats

Graphing a Rational Function

If the numerator or the denominator of a rational function consists of more than one term, you must enclose it in parentheses when you enter it into your calculator.

Example: Graph $r(x) = \dfrac{x^2 - 1}{x}$ using a standard viewing window.

- Press $\boxed{\text{Y=}}$ and erase any previously stored functions. (Position the cursor on a function and press $\boxed{\text{CLEAR}}$.)

- Enter $r(x)$: press $\boxed{(}$ $\boxed{\text{X,T},\theta}$ $\boxed{x^2}$ $\boxed{-}$ $\boxed{1}$ $\boxed{)}$ $\boxed{\div}$ $\boxed{\text{X,T},\theta}$. Then press $\boxed{\text{ZOOM}}$ $\boxed{6}$ to view the graph in the standard viewing window. (Draw a quick sketch of the graph so you remember what it looks like.)

Now, let's see what would happen if you forgot to enclose the numerator of $r(x)$ in parentheses.

- Press $\boxed{\text{Y=}}$.

- Use $\boxed{\text{DEL}}$ to delete the parentheses that surround the numerator.

- Press $\boxed{\text{GRAPH}}$ to view the graph. Did removing the parentheses affect the shape of the graph?

The formula for the function whose graph now appears on your screen is $x^2 - \frac{1}{x}$. Without parentheses, the calculator divides only the last term, 1, by x.

Zooming Out

Example: Graph the function $f(x) = \dfrac{x^2 - 1}{x - 3}$ in the standard viewing window.

Once more with feeling: When you enter a rational function, be sure to enclose both the numerator and denominator in parentheses; if you need help graphing the function in this example, refer to the instructions for the previous example. If you have entered the function correctly, your graph should match the one on the screen below.

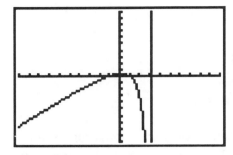

The vertical line in the graph above indicates that this function has a vertical asymptote at $x = 3$. Remember, this line is not part of the graph of the function. Furthermore, because the domain of this function is all real numbers except $x = 3$, there is a branch of this graph that lies to the right of the line $x = 3$. In order to observe this branch, you would have to adjust the WINDOW setting for Ymax. We'll do just that on page 44.

Now, let's see what happens to the appearance of the graph of $f(x)$ as we "back away" by increasing the width and height of the viewing window.

- Press $\boxed{\text{WINDOW}}$ and set Xscl and Yscl to 0. (This turns off the tick marks that appear on the axes. If you skip this step, the axis will get crowded with tick marks when you zoom out.)

- Check the settings for the zoom factor: press $\boxed{\text{ZOOM}}$ and highlight MEMORY. Select $\boxed{4}$ for SetFactors. Your screen should match the one below. If it doesn't, adjust the factor settings before proceeding to the next step.

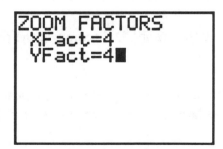

- To view the graph over wider x- and y-intervals, press $\boxed{\text{ZOOM}}$ $\boxed{3}$ for Zoom Out. (A blinking pixel, part of a free moving cursor, should appear in the center of your screen. The zooming will be centered around this location. If you wish to change the focal point of the zooming, use the arrow keys to position this cursor at a different center.)

Chapter 4

- Press ENTER to view the graph over wider x- and y-intervals. Press ENTER again to zoom out a second time. (The graph should look like a line except, perhaps, for a small blip slightly to the right of the origin.)

- Press WINDOW to observe the affect on the WINDOW settings of twice zooming out (by a factor of 4).

The default setting for Zoom Out widens both the x- and y-intervals by a factor of 4 each time that it is applied. In the previous example, you zoomed out twice. Therefore, the x- and y-intervals are 16 times wider than they were before you zoomed out.

In the next example we'll change the default zoom settings in order to observe the behavior of a function that begins to act like its horizontal asymptote.

Example: Graph $q(x) = \dfrac{5x^2 + 20x - 105}{2x^2 + 2x - 60}$ in the standard viewing window.

If you have entered the function correctly, your graph should look like this:

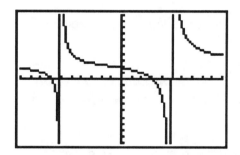

Based on this graph you may suspect that the function $q(x)$ has a horizontal asymptote but it is not at all obvious. Let's observe the function's graph over increasingly wide x-intervals to see if it begins to behave like a horizontal line. Instructions on how to zoom out in the horizontal direction follow.

- Press WINDOW and turn off the tick markings by setting Xscl and Yscl to 0.

- Press ZOOM and highlight MEMORY. Select 4 for SetFactors. To leave the y-interval unchanged and zoom out in the horizontal direction only, set YFact to 1 and XFact to 4.

- Press GRAPH to return to the graph of $q(x)$. Now press ZOOM 3 ENTER to widen the x-interval by a factor of four. Press ENTER several more times to continue widening the x-interval. Your graph should begin to resemble its horizontal asymptote $y = 2.5$.

TI-82 Guide

Project 4.4: Vertical Asymptotes and "Black Holes"

Vertical Asymptotes

Example: Let's return to the graph of $f(x) = \dfrac{x^2 - 1}{x - 3}$ in the standard viewing window.

Graph $f(x)$ in the standard viewing window. Your graph should match the graphing screen on page 42.

For this example, your calculator draws the vertical asymptote at $x = 3$. However, this viewing window does not show any graph to the right of $x = 3$. Since the domain of $f(x)$ includes all real numbers larger than 3, we need to adjust the viewing window in order to see what the graph looks like to the right of the line $x = 3$.

Let's change the viewing window so that we can observe the branch of the graph that lies to the right of the vertical asymptote.

- Press $\boxed{\text{WINDOW}}$ and change Ymax to 20.

- Now press $\boxed{\text{GRAPH}}$ to obtain a more comprehensive view of the graph.

Warning: *In the previous example, the TI-82 drew the vertical asymptote for you. However, in some viewing windows, the calculator does not draw the vertical asymptote. Therefore, you will need to keep track of the vertical asymptotes yourself and not rely exclusively on the calculator.*

Let's change the viewing window to match the one that follows and look at one last graph of $f(x)$.

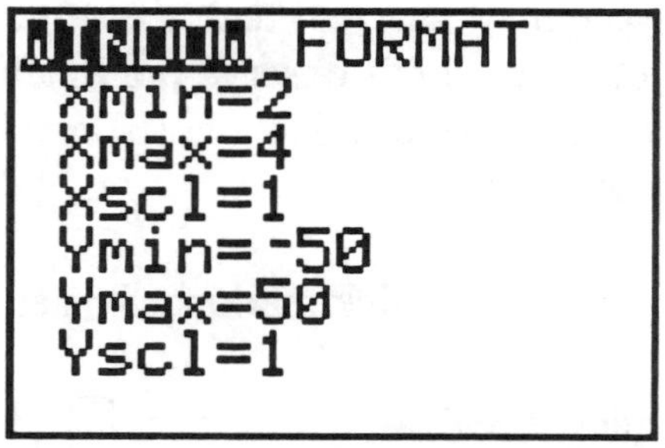

```
WINDOW FORMAT
 Xmin=2
 Xmax=4
 Xscl=1
 Ymin=-50
 Ymax=50
 Yscl=1
```

This time portions of both branches of the graph were visible; however, the TI-82 did not draw the vertical asymptote.

Holes

Example: Graph $h(x) = \dfrac{x^2 - 4}{x - 2}$ in the two viewing windows specified below.

Window 1: Graph $h(x)$ in the standard viewing window.

Notice that $h(x)$ does not have a vertical asymptote! Instead, there is a hole in the graph at $x = 2$ that can't be seen in the present viewing screen.

Window 2: Graph $h(x)$ in the viewing window shown below.

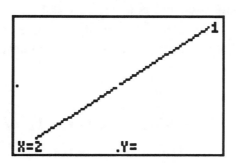

If you look carefully, you can now see the hole in the graph at (2, 4). (One pixel will be missing from the line.) The screen that follows shows what happens when you try to get the coordinates for this hole using TRACE. To see why your calculator gives no value for y, try evaluating $h(2)$ yourself, by hand. What problem do you encounter?

Project 4.5: Long-Term Behavior of Rational Functions

The Zoom Out feature on your calculator can be very useful for this project. See *Zooming Out* on pages 41 − 43 for details.

TI-82 Guide

When the Degree of the Denominator Is Greater Than or Equal To the Degree of the Numerator

Example: Explore the long-term behavior of the rational function $r(x) = \dfrac{6x+5}{2x-3}$.

Plan of action: Since the denominator of this function has the same degree as its numerator, we'll examine its long-term behavior by first graphing $r(x)$ in a viewing window that shows its key local features and then by zooming out in the x-direction only.

- Start by graphing $r(x)$ in the standard viewing window.

- Set the zoom factors: Press ZOOM, highlight MEMORY, and then press 4 for SetFactors. Set XFact to 4 and YFact to 1.

- Press WINDOW and set Xscl to 0. (This will remove the tick marks from the x-axis.)

- Press GRAPH to return to the graph of $r(x)$.

- Press ZOOM 3 ENTER to widen the width of the x-interval by a factor of 4. Then press ENTER a second and then third time to widen the interval by a factor of 16 and 64, respectively.

- Press TRACE and use the right $\rightarrow$ and left $\leftarrow$ arrow keys to move along the graph. (As you trace along the graph, the values of the y-coordinates should be approximately 3 except for points on the graph near the y-axis.)

When the Degree of the Denominator Is Less Than the Degree of the Numerator

Example: Explore the long-term behavior of the rational function $q(x) = \dfrac{4x^2 + x - 6}{x + 4}$.

Plan of action: First, we'll graph $q(x)$ in a viewing window that shows some of its local features. Then, since the degree of the denominator is less than that of the numerator, we'll examine the long-term behavior of $q(x)$ by zooming out in both the x- and y-directions.

- Graph $q(x)$ in the standard viewing window.

Chapter 4

- Set the zoom factors: Press ZOOM, highlight MEMORY, and then press 4 for SetFactors. Set both XFact and YFact to 4.

- Press WINDOW and set Xscl and Yscl to 0. (This will remove the tick marks from the x- and y-axes.)

- Press GRAPH to return to the graph of $r(x)$.

- Press ZOOM 3 ENTER to zoom out. Then press ENTER two more times. Your graph should resemble the line with positive slope shown below.

Chapter 5

General Information

Graphing Exponential Functions

In an exponential function, the independent variable is part of an exponent. You'll need to use the $\boxed{\wedge}$ key to enter the exponent. Also, because exponential functions increase or decrease very quickly in certain regions of their domain, you may have to experiment in order to find a viewing window that captures the function's important graphical features.

Example: Graph $y = 4^x$ in the standard viewing window.

To enter 4^x, press $\boxed{4}\ \boxed{\wedge}\ \boxed{X,T,\theta}$.

Notice that, to the left of the y-axes, the graph appears to merge with the line $y = 0$ and to the right of the y-axes, the graph becomes so steep that it appears vertical.

Let's change the viewing window so that the graph will fill more of the screen. Adjust the RANGE settings to match the ones that follow.

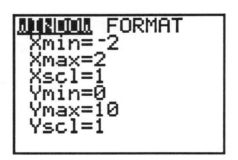

An exponential function has a positive number as its base (the base is the number that gets raised to the exponent). The next example should help you understand why we don't deal with negative bases.

Example: Graph $y = (-4)^x$ in the standard viewing window.

To enter $(-4)^x$, press $\boxed{(}\ \boxed{(-)}\ \boxed{4}\ \boxed{)}\ \boxed{\wedge}\ \boxed{X,T,\theta}$.

For this function your calculator displays a blank graph. If you adjust the WINDOW settings, you will still get a blank graph. To see why, try evaluating $y = (-4)^x$ for several x-values, say $x = .5,\ 1,\ 1.25,\ 1.5,$ and 2. Which of these x-values produce real number outputs?

There are two bases for exponential functions, 10 and e, that are so common they have their own function keys on the calculator, $\boxed{[10^x]}$ and $\boxed{[e^x]}$.

Example: Graph $f(x) = e^x$ and $g(x) = 10^x$.

- Press $\boxed{Y=}$. Erase any previously stored functions by positioning the cursor on the function and pressing $\boxed{\text{CLEAR}}$.

- To enter $f(x)$, press: $\boxed{\text{2nd}}$ $\boxed{[e^x]}$ (same key as $\boxed{\text{LN}}$) $\boxed{X,T,\theta}$.

- To enter $g(x)$, press: $\boxed{\text{2nd}}$ $\boxed{[10^x]}$ (same key as $\boxed{\text{LOG}}$) $\boxed{X,T,\theta}$.

- Press $\boxed{\text{WINDOW}}$ and adjust the settings to match the ones below.

```
WINDOW FORMAT
Xmin=-2
Xmax=3
Xscl=0
Ymin=-1
Ymax=12
Yscl=0
```

- Press $\boxed{\text{GRAPH}}$ to view the graphs.

Graphing Logarithmic Functions

The logarithmic functions with base e and base 10 have their own function keys $\boxed{\text{LN}}$ and $\boxed{\text{LOG}}$ on the calculator. Logarithmic functions of other bases can be graphed by dividing these functions by the appropriate scaling factor. (Refer to the Algebra Appendix, A.4(ii).)

Example: Graph $F(x) = \ln(x)$ and $G(x) = \log(x)$.

- Press $\boxed{Y=}$. Erase any previously stored functions by positioning the cursor on the function and pressing $\boxed{\text{CLEAR}}$.

TI-82 Guide

- To enter F(x): press [LN] [X,T,θ].

- To enter G(x): press [LOG] [X,T,θ].

- Press [WINDOW] and adjust the settings to match the ones below.

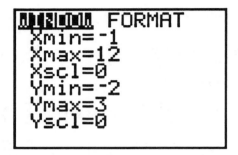

- Press [GRAPH] to view the graphs.

Lab 5A: AIDS

Graphing an Exponential Function Expressed in Base-e Form

Example: Graph $y = 3e^{.5x}$ in the standard viewing window.

- Press [Y=]. Erase any stored functions.

- To enter the function, press [3] [2nd] [e^x] [.] [5] [X,T,θ].

- Press [ZOOM] [6] to view the graph in the standard viewing window.

Unlike many computer programs such as MAPLE, the TI-82 is not fussy about whether or not you enclose the exponent, $.5x$, in parenthesis. However, if you enter the exponent as $.5*x$, using the multiplication key, then you must enclose the exponent in parenthesis. Here is an example of what you cannot do:

- Press [Y=] and enter [2nd] [e^x] [.] [5] [×] [X,T,θ].

- Press [GRAPH] to view the graph.

Although you may have expected an exponential function, your graph represents the linear function $y = (e^{.5})x$.

50

Chapter 5

Warning! *You must take care when interpreting the calculator-produced graphs of exponential functions. In places the graph is so steep that you might think that the function has a vertical asymptote. In other places the function is so close to zero that the graph in your viewing window merges with the x-axis, even though the function is never zero.*

Lab 5B: Radioactive Decay

Graphing a Function Representing Exponential Decay

Example: Graph $y = e^{-\frac{x}{2}}$ in the standard viewing window.

There are two things that you must remember when entering this function. Use the $\boxed{(-)}$ key for the opposite of $\frac{x}{2}$ and enclose the exponent, $-\frac{x}{2}$, in parentheses. Your graph should match the one below.

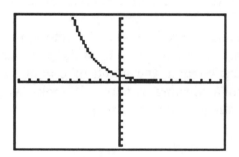

Lab 5C: Earthquakes

Tracing Beyond the Viewing Window

As with exponential functions, it is difficult to get a comprehensive picture of logarithmic functions from a single viewing window. For x-values near zero, the graph of a logarithmic function is so steep that it will appear to merge with the y-axis. As x-values become large, the graph becomes so flat that you might think that the function has a horizontal asymptote.

Example: Trace the x- and y-values on the graph of $y = \ln(x)$ beyond the viewing window.

- Clear any previously stored functions.

TI-82 Guide

- Press [Y=] [LN] [X,T,θ].

- Press [WINDOW]. Change the window settings to match the ones shown below.

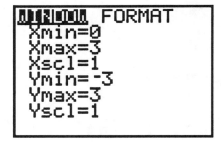

- Press [TRACE] and then press down the right arrow key [→]. As you continue to hold down the right arrow key [→], the window settings will automatically change to let you trace along the portion of the graph that lies outside the viewing window. Watch how slowly the y-values increase as the x-values increase. (Release the right arrow key from time to time so that you can read off x- and y-values.)

Remember that $\ln(x)$ and e^x are inverses. Therefore, the range of $\ln(x)$ is the same as the domain of e^x, the set of real numbers. If, in the previous example, you press down on the right arrow key long enough, the graph of $y = \ln(x)$ will look constant in the viewing window. In such a viewing window, it looks as if $\ln(x)$ has a horizontal asymptote when, in fact, its range is the set of real numbers.

Chapter 6

General Information

Graphing Trigonometric Functions

Three of the six basic trigonometric functions are built-in functions on the TI-82: sine [SIN], cosine [COS], and tangent [TAN]. Before graphing any of these functions, you should first check that your calculator is set in radian mode. Press [MODE] and highlight **Radian** if it is not already highlighted.

The standard window is generally not the best window to use when graphing trigonometric functions. Your calculator's trigonometric viewing window (ZTrig) frequently provides a better scale to start with.

Example: Graph $y = \sin(x)$ in the trigonometric viewing window.

- Press [Y=] [SIN] [X,T,θ].

- To set the axes for the trig functions, press [ZOOM] [7]. Your graph should match the one that follows.

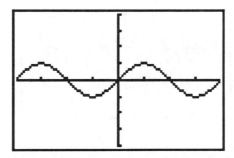

- Press [WINDOW] to observe the settings for the trig viewing window.

Problems Inherent in the Technology
(Don't Believe Everything That You See!)

Your viewing screen consists of a grid of pixels. (If you darken the contrast to its maximum setting, you may be able to see the grid.) When a pixel is *on*, it shows up as a

TI-82 Guide

dark square dot on the screen. Graphs are formed by turning on a series of pixels. This method of producing graphs can sometimes produce misleading images.

Let's look at what happens when we graph $y = \sin(x)$ over increasingly wide x-intervals.

- Start with a graph of sin(x) in the trigonometric viewing window. (Refer to the previous example.)

- Turn off the tick marks for the x-axis: press WINDOW and set Xscl to 0.

- Change the Zoom factors: press ZOOM , highlight MEMORY, and press 4 . Set XFact to 10 and YFact to 1.

- Press Graph .

Get ready for some fun. When sin(x) is graphed in the trig viewing window, you can observe two complete cycles (from -2π to 2π) of the wave. Each time you increase the width of the x-interval by a factor of ten, you should see ten times as many cycles of the sine wave.

- Press ZOOM 3 ENTER . Count the number of complete cycles of the sine wave that appear in this viewing window. (You should see approximately 20 cycles.)

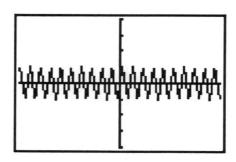

In this case, the calculator shows the correct number of cycles. However, because graphs on the calculator consist of a discrete set of highlighted pixels, your calculator is unable to produce the smooth wavy curve that is characteristic of sine waves. If you press enter again, you should expect ten times as many cycles as in the picture above, that is, around 200 cycles of sin(x).

- Now press ENTER again to see what actually appears.

This time the graph looks smooth with far fewer cycles than in the picture above! Here's what's going on. In producing this picture, your calculator does not have enough pixels to capture all the oscillations that are part of the actual graph. The small subset of

54

points from the actual graph that the calculator chooses to represent with darkened pixels present a very misleading picture of the features of the actual graph!

Lab 6: Daylight and SAD

Modifying Basic Trigonometric Functions

Example: Graph $g(x) = \cos\left(x + \frac{\pi}{2}\right)$ in the trig viewing window.

To enter $\cos\left(x + \frac{\pi}{2}\right)$, press [COS] [(] [X,T,$\theta$] [+] [2nd] [$\pi$] (same key as [^]) [÷] [2] [)].

The trigonometric viewing window gives a good picture of g. However, it is not the best window for viewing all trigonometric functions. Let's look at another example and use our understanding of how the constants in a trigonometric function affect its graph to help select a good viewing window.

Example: Graph $y = 5\cos(10x)$.

- First, look at the function in the trig viewing window.

The trig viewing window is not very good for displaying the key features of the function, so, we'll need to adjust the WINDOW settings. Let's think about the effect that the constants 5 and 10 have on the basic cosine function. Recall from Projects 3.1 and 3.2 that multiplying a function by 5 stretches the basic graph vertically by a factor of 5 and that multiplying the input by 10 compresses the basic graph horizontally by a factor of 10. Thus, you will get a good viewing window for this function if you adjust the WINDOW settings as follows: divide the x settings by 10 and multiply the y settings by 5. Try it!

- Press [WINDOW]. Change the x-settings to one-tenth of their present values and the y-settings to five times their present values.

- Press [GRAPH] to observe the graph.

- Now press [TRACE] for TRACE. Use the right and left arrow keys to trace along the curve. What is the amplitude of this function?

Project 6.4: Don't *Lean* On Me

Away Dull Trig Tables!

With the TI-82 you can solve problems in right-triangle trigonometry without using trig tables. To compute the sine, cosine, or tangent of an angle measured in degrees, first change the Radian/Degree mode setting to Degree: press $\boxed{\text{MODE}}$ and highlight Degree by moving the cursor over Degree and pressing $\boxed{\text{ENTER}}$.

Example: Compute sin 30° and tan 25 °.

- Press $\boxed{\text{SIN}}$ $\boxed{3}$ $\boxed{0}$ $\boxed{\text{ENTER}}$. (If you do not get .5 for the answer, go back and check that you have changed your calculator to degree mode.)

- Check that tan (25 °) ≈ .466.

Example: Compute $\cos^{-1}(.5)$ and $\tan^{-1}(2.5)$.

- Press $\boxed{\text{2nd}}$ $\boxed{[\text{COS}^{-1}]}$ $\boxed{.}$ $\boxed{5}$ $\boxed{\text{ENTER}}$. Did you get approximately 60? Remember this is 60° if you're in degree mode.

- Check that $\tan^{-1}(2.5)$ ≈ 68.199°.

Finally, here's a way to compute the sine, cosine, or tangent of an angle given in degrees without changing the mode setting. First, return your calculator to its default mode setting, Radian: press $\boxed{\text{MODE}}$, highlight Radian, and press $\boxed{\text{ENTER}}$. Now, let's compute cos (60°) and sin (35°) without changing the mode setting. (This is a good idea because for most of your work in precalculus, you want radian mode.)

Example: Compute cos 60° and sin 35°.

- Press $\boxed{\text{COS}}$ $\boxed{6}$ $\boxed{0}$ $\boxed{\text{2nd}}$ $\boxed{[\text{ANGLE}]}$ $\boxed{1}$ $\boxed{\text{ENTER}}$. You should get .5 for your answer.

- Now use this method for sin (35°). Did you get approximately .574?

Chapter 7

General Information

Graphing Parametric Equations

For graphing parametric equations, you'll need to change your calculator from function (Func) mode to parametric (Par) mode. Here's how: press $\boxed{\text{MODE}}$, then use the arrow keys to select **Par** and press $\boxed{\text{ENTER}}$.

Now let's see the changes in the Y= and WINDOW menus.

- Press $\boxed{\text{Y=}}$. You can enter six sets of equations, XT and YT. (If you have any functions stored in this menu, erase them by positioning the cursor on a function and pressing $\boxed{\text{CLEAR}}$.)

- Press $\boxed{\text{ZOOM}}$ $\boxed{6}$ to set up the standard viewing window for parametric equations.

- Now press $\boxed{\text{WINDOW}}$ to observe the WINDOW settings. Use the up $\boxed{\uparrow}$ and down $\boxed{\downarrow}$ arrow keys to scroll through the entries in this menu. Notice that the settings associated with x and y are the same as they are in function mode. However, when your calculator is in parametric mode you must also specify bounds and increments for the parameter t: Tmin, Tmax, Tstep.

For example, suppose we want to graph the set of parametric equations

$$x(t) = 2t + 1, \qquad y(t) = -3t + 5$$

in the standard viewing window.

- Press $\boxed{\text{Y=}}$. (Erase any previously stored parametric equations by positioning the cursor on the line containing the function and pressing $\boxed{\text{CLEAR}}$.)

- Enter the equation for x opposite X1T: press $\boxed{2}$ $\boxed{\text{X,T,}\theta}$ $\boxed{+}$ $\boxed{1}$. (Notice that in this mode, pressing $\boxed{\text{X,T,}\theta}$ enters a T.)

- Enter the equation for y opposite Y1T: press $\boxed{\text{(-)}}$ $\boxed{3}$ $\boxed{\text{X,T,}\theta}$ $\boxed{+}$ $\boxed{5}$.

- Press ZOOM 6 to view the graph in the standard viewing window. You should see a graph similar to the one below.

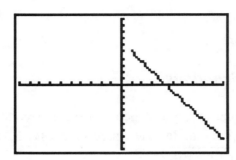

- Press TRACE. The cursor will mark the location when $t = 0$. The value of t and the coordinates of the point will appear at the bottom of your screen. Press the right arrow key → and the cursor will jump to the location associated with $t \approx .13$ (t increases by one Tstep). Press → repeatedly and watch the cursor move along the line.

Next, let's graph the position of a dot as it moves along the path

$$x(t) = 2t + 1, y(t) = -3t + 5$$

at .5 second increments from time $t = 0$ seconds to $t = 3$ seconds.

For this example, we assume that your calculator is in parametric mode and that you have already entered this set of parametric equations in your calculator from the previous example.

- Press MODE and highlight Dot.

- Press WINDOW. Adjust the parameter settings for t to match those on the screen that follows.

Chapter 7

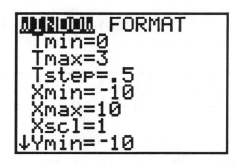

- Press TRACE . Then press → and watch the dot move from one position to the next (on a screen similar to the one below) in .5 second time increments.

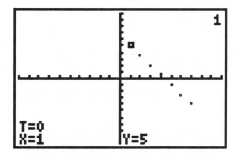

Lab 7A: Bordeaux

Graphing a Function of More Than One Input Variable

You can represent functions with more than one input variable graphically by replacing one (or more) of the independent variables with a list of values. (If you don't remember how to handle lists, refer to *Graphing a Family of Functions* on page 34.)

If you changed your calculator from the default mode settings, press MODE and change the settings to the match those in the mode menu on page 5.

Example: Examine the behavior of $F(w, x) = 3w - 2x$ by holding w constant, first at -2, then at 0, and then at 2.

The three graphs, taken together, show how F varies with x for three different values of w. By forming a list $\{-2, 0, 2\}$ of the constant values for w, you can produce the graphs $y = F(-2, x)$, $y = F(0, x)$ and $y = F(2, x)$ using a single functional expression. Here's how:

TI-82 Guide

- Press ZOOM 6, for the standard viewing window. Press Y= and erase (or turn off) any previously stored functions.

- Press 3 2nd [{] (-) 2 , 0 , 2 2nd [}] − 2 X,T,θ. Notice that the input variable w has been replaced by the list of values $\{-2, 0, 2\}$.

- Press GRAPH and watch as three parallel lines are graphed one by one.

Lab 7B: Bézier Curves

Combining Two Sets of Parametric Equations

In Lab 7B you are asked to form a new set of parametric equations from a combination of two other sets of parametric equations.

For example, suppose that you have two sets of parametric equations,

$$S_1: \quad x_1 = 2t + 1$$
$$y_1 = -3t + 5$$

$$S_2: \quad x_2 = t - 5$$
$$y_2 = 4t - 3,$$

and that you want to graph $(1-t)S_1 + t\, S_2$, a combination of these equations, over the interval $0 \leq t \leq 1$.

Step 1: Change your calculator to parametric mode: press MODE, select **Par** and press ENTER.

Step 2: Enter the two sets of parametric equations, S_1 and S_2.

- Erase all previously stored functions: press 2nd [MEM] 2 5 and then press ENTER repeatedly until all functions have been deleted.

- Press Y=.

- Enter S_1's equations as X1T and Y1T. (Refer to *Graphing Parametric Equations*, page 57, if you have trouble.)

Chapter 7

- Enter S$_2$'s equations as X2T and Y2T.

Step 3: Enter the x and y equations for the combination, $(1-t)$S$_1 + t$ S$_2$:

$$x_3 = (1-t)x_1 + tx_2$$
$$y_3 = (1-t)y_1 + ty_2$$

- The cursor should be opposite X3T. Enter the x-equation for the combination:
  Press $\boxed{(}$ $\boxed{1}$ $\boxed{-}$ $\boxed{\text{X,T},\theta}$ $\boxed{)}$.
  Press $\boxed{\text{2nd}}$ $\boxed{\text{Y-VARS}}$ $\boxed{2}$ for Parametric, $\boxed{1}$ for X1T.
  Press $\boxed{+}$ $\boxed{\text{X,T},\theta}$ $\boxed{\text{2nd}}$ $\boxed{\text{Y-VARS}}$ $\boxed{2}$ $\boxed{3}$ for X2T.

- Enter the y-equation for the combination: move the cursor opposite Y3T.
  Press $\boxed{(}$ $\boxed{1}$ $\boxed{-}$ $\boxed{\text{X,T},\theta}$ $\boxed{)}$.
  Press $\boxed{\text{2nd}}$ $\boxed{\text{Y-VARS}}$ $\boxed{2}$ $\boxed{2}$ for Y1T.
  Press $\boxed{+}$ $\boxed{\text{X,T},\theta}$ $\boxed{\text{2nd}}$ $\boxed{\text{Y-VARS}}$ $\boxed{2}$ $\boxed{4}$ for Y2T.

Step 4: Graph the combination, $(1-t)$S$_1 + t$ S$_2$.

- Turn off the parametric equations for X1T, X2T, Y1T, and Y2T. Move the cursor over X1T's equals sign and press $\boxed{\text{ENTER}}$. Next, move the cursor over X2T's equals sign and press $\boxed{\text{ENTER}}$. (Notice the highlighting has also been removed from Y1T and Y2T's equals signs.)

- Press $\boxed{\text{MODE}}$ and highlight Connected.

- Press $\boxed{\text{WINDOW}}$. Set Tmin = 0 and Tmax = 1. Adjust the remainder of the settings to match the ones below.

```
WINDOW FORMAT
 Tstep=.1
 Xmin=-5
 Xmax=2
 Xscl=1
 Ymin=-4
 Ymax=6
 Yscl=1
```

- Press GRAPH. Your graph should look like this:

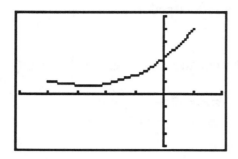

Project 7.3: What Goes Around Comes Around

Using Square Scaling in Parametric Mode

Press MODE and check that Radian, Par, Dot, and Sequential are all highlighted.

In addition to thinking about the x- and y-settings for the viewing window, you will need to adjust the WINDOW settings for the parameter t. Use the following t-settings for all questions except question 8: Set Tmin to 0, Tmax to 6.3, and Tstep to .1.

You will want to use square scaling for all your viewing windows.

- First, select a viewing window that shows the basic details of your graph.

- Then press ZOOM 5 to adjust to square scaling.

Example: Graph the set of parametric equations $x = 3\cos(t)$, $y = 3\sin(t) + 2$.

- Press MODE and check that your settings match the ones below.

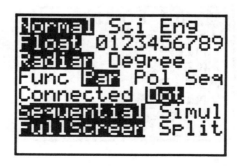

Chapter 7

- Clear any previously stored functions: press [2nd] [MEM] [2] for Delete, and then press [5] for Y-Vars. Press [ENTER] repeatedly until all functions are erased.

- Press [Y=] and enter the set of parametric functions.

- Press [Window]. Set Tmin to 0, Tmax to 6.3, and Tstep to .1. Adjust the remaining settings to match the those on the following screen.

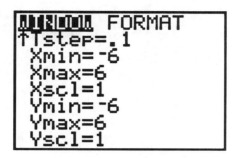

- Press [GRAPH]. Your graph should look egg-shaped. Next, press [ZOOM] [5] to observe the graph in a window with square scaling. Your graph should resemble the one below.

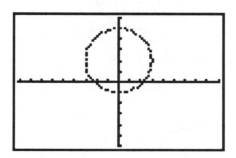

Basic Tutorial for the TI-85

As you proceed through this guide, note that specific keys that you are to press appear in a box. For example, you may be asked to press $\boxed{\text{Graph}}$. Operations corresponding to the blue or yellow lettering above the keys are indicated in brackets, $\boxed{[]}$. To access yellow upper key functions, press $\boxed{\text{2nd}}$ followed by the key. To access blue upper key functions, press $\boxed{\text{ALPHA}}$ and then the key.

1. *On, Off, and Contrast*

Turn the calculator on by pressing $\boxed{\text{ON}}$.

You may need to adjust the contrast. Press $\boxed{\text{2nd}}$ followed by holding down the up arrow key $\boxed{\uparrow}$ to darken or the down arrow key $\boxed{\downarrow}$ to lighten. (The four arrow keys are located directly below the $\boxed{\text{F4}}$ and $\boxed{\text{F5}}$ keys.)

To turn your calculator off, press $\boxed{\text{2nd}}$ $\boxed{\text{[OFF]}}$. If you forget, the calculator will automatically turn off after a period of non-use.

2. *Basic Calculation and Editing*

The screen that displays your calculations is called the Home screen. If your calculator is in a menu or displays a graph when it is turned off, it will return to the menu or graph when you turn it back on. Press $\boxed{\text{EXIT}}$ one or more times, or press $\boxed{\text{2nd}}$ $\boxed{\text{[QUIT]}}$, to return to the Home screen.

At times, you'll want to start with a clear Home screen. To remove previous calculations, press $\boxed{\text{CLEAR}}$. You do not have to clear the Home screen after each computation.

Example: Compute 3 × 4.
  After pressing $\boxed{3}$ $\boxed{\times}$ $\boxed{4}$, press $\boxed{\text{ENTER}}$. Note that the original problem, written as 3 ∗ 4, remains on the left side of the screen and the answer appears to the right.

Example: Compute $3 + 2 \times 6$ and $(3 + 2) \times 6$.

In which order did the calculator perform the operations of addition and multiplication?

Example: Compute 8^2 and 1.05^7.

Press $\boxed{8}$ followed by $\boxed{x^2}$ $\boxed{\text{ENTER}}$. You can also compute the square of eight by pressing $\boxed{8}$ $\boxed{\wedge}$ $\boxed{2}$. Now try 1.05^7 using $\boxed{\wedge}$ $\boxed{7}$ to create the power.

Example: Compute $\sqrt{16}$.

Press $\boxed{\text{2nd}}$ $\boxed{[\sqrt{}]}$ (same key as $\boxed{x^2}$) followed by $\boxed{1}$ $\boxed{6}$ $\boxed{\text{ENTER}}$.

Example: Compute $\sqrt{-16}$.

Press $\boxed{\text{2nd}}$ $\boxed{[\sqrt{}]}$ followed by $\boxed{(-)}$ (the key to the left of $\boxed{\text{ENTER}}$) $\boxed{1}$ $\boxed{6}$ $\boxed{\text{ENTER}}$.

Many calculators would report an error message when you tried to compute $\sqrt{-16}$ because there is no real number whose square is -16. However, your TI-85 calculator, unable to compute $\sqrt{-16}$ using real numbers, does the computation using complex numbers. Your TI-85 responds: (0, 4), which represents the number $0 + 4 \cdot \sqrt{-1}$

Example: Compute $\sqrt[5]{32}$ and $\sqrt[3]{18}$.

To compute $\sqrt[5]{32}$:

• Press $\boxed{5}$ for the fifth root.

• Press $\boxed{\text{2nd}}$ $\boxed{[\text{MATH}]}$ (same key as $\boxed{\times}$) to access the MATH menu, then press $\boxed{\text{F5}}$ to choose MISC. Your screen should match the one below.

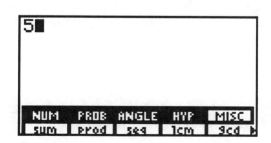

65

TI-85 Guide

The first five choices of the MATH/MISC menu are displayed on the bottom line of the menu.
- Press $\boxed{\text{MORE}}$ to reveal more options. The root operation, $\sqrt[x]{\ }$, will now appear above the $\boxed{\text{F4}}$ key. Press $\boxed{\text{F4}}$ to select $\sqrt[x]{\ }$.

- Press $\boxed{3}$ $\boxed{2}$ $\boxed{\text{ENTER}}$.

<u>Before</u> exiting this menu, let's approximate $\sqrt[3]{18}$. Here's how:

- Since you are already in the MATH/MISC menu, press $\boxed{3}$, for the cube root, and then $\boxed{\text{F4}}$ $\boxed{1}$ $\boxed{8}$ $\boxed{\text{ENTER}}$.

- Now you can exit the MATH/MISC menu: press $\boxed{\text{EXIT}}$ twice (the first time to exit MISC and the second, to exit MATH).

Warning: *The TI-85 has two minus keys, $\boxed{-}$ and $\boxed{(\text{-})}$, to differentiate between the operation of subtraction (such as $3 - 2 = 1$) and the opposite of the positive number 2, namely, -2.*

Example: Compute $-2 + 5$
    To compute $-2 + 5$, press $\boxed{(\text{-})}$ then $\boxed{2}$ to create the -2. Finish the computation to get the answer 3.

3. *Correcting an Error*

We tackle three situations connected with making errors. First, we look at two examples of errors that your calculator recognizes as errors. Then we provide an example that illustrates what you can do when you discover that you have punched in an error that the calculator is able to compute.

Correcting by Deleting

Let's start by making a deliberate error: press $\boxed{3}$ $\boxed{+}$ $\boxed{+}$ $\boxed{2}$ $\boxed{\text{ENTER}}$.
    The following message will appear on your screen.

Basic Tutorial

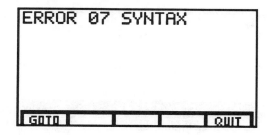

Press F1 and the cursor will GOTO the error. Erase one of the plus signs by pressing DEL for delete, and then ENTER. The correct answer to 3 + 2 will appear.

Correcting by Starting Over

Try entering √2 as 2 2nd [√] ENTER.
To correct your error, press F5 to QUIT. Then press CLEAR to clear the screen, and you can start over.

Correcting by Inserting

Finally, press 3 + 4 ENTER and suppose that you really wanted − 33 + 4. Press 2nd [ENTRY] to return to the previous command. Use the left arrow key ← to position the cursor over the 3. Press 2nd [INS]. (Notice that the cursor changes from a box to a line.) Now press (-) 3 ENTER.

4. *Resetting the Memory*

Warning: *Resetting the memory erases all data and programs. All calculator settings will return to the default settings.*

Here's how to reset your calculator:

- Press 2nd [MEM] F3 to access the MEM/RESET menu that follows.

TI-85 Guide

MEM clears the memory (programs, data, functions, graphs) but leaves the mode settings alone, DFLTS leaves the memory alone but returns the mode settings to their defaults, and ALL resets both memory and defaults.

- (If you have programs or data that you do not want to erase, skip the step that follows. Instead, press F3 to return only the mode settings to the defaults.) Press F1 to reset the memory and return all calculator settings to the defaults. The calculator will respond: **Are you sure?** Press F4 for YES. The message **Mem cleared, Defaults set** should appear on your screen. If you can't read this message, darken the contrast by pressing 2nd and then holding down the up arrow key ↑.

- Press CLEAR to clear the screen.

5. *Changing the Mode Settings*

Press 2nd [MODE]. The cursor should be blinking on NORMAL. If you have reset your calculator as shown in Topic 4, the mode settings will match the default settings shown below.

Suppose that you want all your answers to be displayed in scientific notation. Use the right arrow key → to highlight **Sci** (for scientific notation) and then press ENTER. Press EXIT to exit the mode menu.

Example: Compute 3654×1781.
  The displayed answer should read: $6.507774 \, \text{E} \, 6$. (See the Algebra Appendix, Section A.3(ii) for an explanation of this notation.)

Basic Tutorial

To return the mode setting to NORMAL, press [2nd] [MODE]; The cursor should be blinking on NORMAL, so, press [ENTER]. (Don't exit the MODE menu yet.)

Next use the down arrow key [↓] to move to the second line of the mode menu. Using the right arrow key [→], move the cursor so that it is blinking over the number 2. Press [ENTER] followed by [EXIT].

Example: Compute $3.542 + 2.007$.
Did the calculator truncate or round when it displayed the answer to two-decimal-place accuracy?

Press [2nd] [MODE], position the cursor over FLOAT, and press [ENTER] to return to the default decimal setting. Press [EXIT] to exit the MODE menu.

6. Graphing

Press [2nd] [MODE]. If necessary, adjust settings to match the default settings shown in Topic 5. If you did not reset your calculator in Topic 4, you should erase any previously stored functions. (See Topic 7, *Erasing Stored Functions* (page 73), for instructions.)

Set the standard viewing window:

- Press [GRAPH] to enter the GRAPH menu, [F3] for the GRAPH/ZOOM menu, and then press [F4] for ZSTD. An empty viewing window will appear on your screen. The GRAPH/ZOOM menu will remain at the bottom of your screen.

- Press [EXIT] to return to the GRAPH menu from the GRAPH/ZOOM menu. Now press [F2] for RANGE. The settings for the standard viewing window (shown below) should appear on your screen.

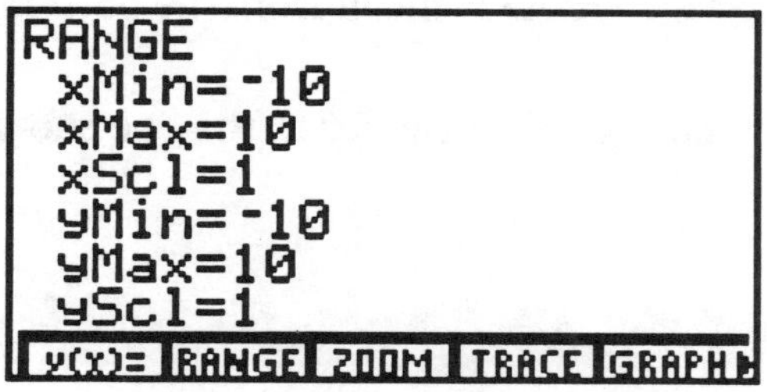

```
RANGE
 xMin=-10
 xMax=10
 xScl=1
 yMin=-10
 yMax=10
 yScl=1
y(x)= RANGE ZOOM TRACE GRAPH▶
```

TI-85 Guide

Notice that the scaling on the x- and y-axes goes from -10 to 10 with tick marks one unit apart (since xScl and yScl = 1).

- Press $\boxed{\text{EXIT}}$. Notice that pressing $\boxed{\text{EXIT}}$ while in the GRAPH/RANGE menu returns you to the Home screen instead of to the GRAPH menu.

Example: Graph $y = x$, $y = x^2$, and $y = x^3$ in the same viewing window.

Step 1: Enter $y = x$.

- Press $\boxed{\text{GRAPH}}$ $\boxed{\text{F1}}$ for the GRAPH/$y(x)$= menu. The cursor will be blinking opposite $y1$.

- Press $\boxed{\text{F1}}$ for x.

- Next, press $\boxed{\text{EXIT}}$ to return to the GRAPH menu. Press $\boxed{\text{F5}}$ for GRAPH. (A pulsating line segment in the upper right corner tells you that the calculator is working on the problem. When the graph is completed, the line segment will disappear. If you missed seeing the pulsating line segment, watch for it on the next graph.)

- Press $\boxed{\text{CLEAR}}$ to clear the menu from the bottom of your screen.

Step 2: Enter the two remaining functions.

- Press $\boxed{\text{GRAPH}}$ $\boxed{\text{F1}}$ $\boxed{\downarrow}$.

- To enter $y = x^2$ as $y2$, press $\boxed{\text{F1}}$ $\boxed{x^2}$ $\boxed{\downarrow}$.

- To enter $y = x^3$ as $y3$, press $\boxed{\text{F1}}$ $\boxed{\wedge}$ $\boxed{3}$.

- Press $\boxed{\text{EXIT}}$ to return to the GRAPH menu, and then $\boxed{\text{F5}}$. Again, press $\boxed{\text{CLEAR}}$ to clear the GRAPH menu from the bottom of the viewing screen.

The picture on your screen should be similar to the one shown that follows.

Basic Tutorial

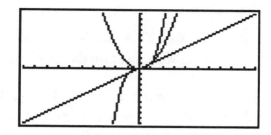

Example: Let's use the TI-85 to find where the linear function, $y = x$, and the cubic function, $y = x^3$, intersect.

Plan of action: Notice that the three points of intersection are difficult to see in the viewing screen shown above. By turning off the function $y = x^2$ and narrowing the viewing window, you'll be able to see the points of intersection more clearly. The instructions for doing this are given in Steps 1 and 2.

Step 1: Remove the graph of $y = x^2$ from the viewing screen.

- Press $\boxed{\text{GRAPH}}$ $\boxed{\text{F1}}$ to display the stored functions.

- Press the down arrow key $\boxed{\downarrow}$ to position the cursor opposite $y2$. Press $\boxed{\text{F5}}$ for SELCT (short for select). This removes the highlighting over $y2$'s equals sign and turns the function off.

- Now, press $\boxed{\text{EXIT}}$ to return to the GRAPH menu. Press $\boxed{\text{F5}}$ $\boxed{\text{CLEAR}}$ to view the graphs.

Note: The graph $y2 = x^2$ no longer appears in the viewing window. You can, however, turn y2 back on by repeating the previous two steps.

Step 2: Magnify the view around the center of the viewing screen.

- Press $\boxed{\text{GRAPH}}$ $\boxed{\text{F3}}$ to enter the GRAPH/ZOOM menu.

- Press $\boxed{\text{F2}}$ for ZIN, the zoom-in option. (Look for a blinking pixel at the center of your viewing window. The coordinate represented by this pixel will remain the center of the viewing window after you have zoomed in.) Press $\boxed{\text{ENTER}}$.

Step 3: Estimate the points of intersection.

Plan of action: First get acquainted with the TRACE feature on the TI-85 and then use TRACE to estimate the points of intersection.

TI-85 Guide

- Press [GRAPH] [F4], for TRACE. The cursor, a box with a blinking × through the diagonals, will appear on the graph of the line $y = x$.

- Press [←] and watch the cursor move to the left along the line. The x- and y-coordinates of the cursor's location will appear at the bottom of the screen. Now press [→] to move the cursor along the line in the opposite direction.

- Press [↑] or [↓] to jump back and forth between the line and the curve.

- Now you are ready to approximate the points of intersection. Position the cursor over one of the points of intersection. Read off the x- and y-values that correspond to the cursor's location. Repeat this process for the other two points of intersection.

How did ZIN affect the window settings? Before zooming in, the viewing window displayed an x-axis scaled from -10 to 10, a width of 20 units. Press [GRAPH] [F2]. Notice that the width of the x-scale has been reduced by a factor of 4.

Example: Graph $y = x^3$ over the x-interval from -10 to 10.

Plan of action: The aim here is to set the y-scale so that the graph remains in the viewing window over the entire x-interval from -10 to 10. Since y has value -1000 when x is -10 and value 1000 when x is 10, you'll need to set yMin and yMax to -1000 and 1000, respectively.

- Return to the GRAPH/$y(x) =$ menu. The cursor should be blinking opposite $y1$. Turn off $y1$: press [F5] to remove the highlighting over $y1$'s equals sign. Then press [EXIT] to return to the GRAPH menu.

- In the GRAPH menu, press [F2] for RANGE. The cursor should be blinking opposite xMin. Enter -10 by pressing [(-)] [1] [0].

- Press [ENTER] or [↓] to position the cursor opposite xMax. Continue changing the settings until your screen matches the one that follows.

Basic Tutorial

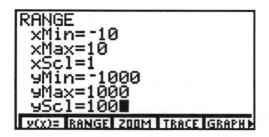

- Now press ⎡F5⎤ for GRAPH.

Note: *The tick marks will be one unit apart on the x-axis and 100 units apart on the y-axis.*

7. Erasing Stored Functions

Now let's see how to erase functions that have been stored in your calculator's memory. After following the instructions in Topic 6, the functions $y = x$, $y = x^2$, and $y = x^3$ should be stored as $y1$, $y2$, and $y3$.

Erase $y = x$, $y = x^2$, and $y = x^3$ from your calculator's memory as follows.

- Return to the GRAPH/$y(x)$= menu (⎡GRAPH⎤ ⎡F1⎤ will do the trick). The cursor should be blinking opposite $y1$.

- Press ⎡CLEAR⎤ to erase the function stored as $y1$.

- Press ⎡↓⎤ to move the cursor opposite $y2$, and then press ⎡CLEAR⎤.

- Press ⎡↓⎤⎡CLEAR⎤ to erase the function stored as $y3$.

- Press ⎡2nd⎤ ⎡[QUIT]⎤ to return to the Home screen.

8. Exiting the Graphing Window or a Menu

Generally, pressing ⎡2nd⎤ ⎡QUIT⎤ or ⎡EXIT⎤ one or more times is all that is needed to return to the Home screen from a menu or viewing window. Each time that you press ⎡EXIT⎤, your calculator will return to the previous menu, display, or the Home screen. When you press ⎡2nd⎤ ⎡QUIT⎤ your calculator will return directly to the Home screen. Two examples designed to give you some practice returning to the Home screen follow.

TI-85 Guide

Example: Return to the Home screen from the GRAPH/$y(x)$= menu.

- Press $\boxed{\text{GRAPH}}$ $\boxed{\text{F1}}$ to enter the GRAPH/$y(x)$= menu. Press $\boxed{\text{2nd}}$ $\boxed{\text{[QUIT]}}$ to return directly to the Home screen.

- Again, press $\boxed{\text{GRAPH}}$ $\boxed{\text{F1}}$. This time press $\boxed{\text{EXIT}}$ twice, the first time to return to the GRAPH menu and the second time to return to the Home screen.

Example: Graph the function $y = 2x$. Then, return to the Home screen.

- Press $\boxed{\text{GRAPH}}$ $\boxed{\text{F1}}$ to enter the GRAPH/$y(x)$ = menu. After clearing any previously stored functions, enter the function $y = 2x$. Press $\boxed{\text{EXIT}}$ to return to the GRAPH menu and then $\boxed{\text{F3}}$ $\boxed{\text{F4}}$ to graph the function in the standard viewing window.

- Press $\boxed{\text{CLEAR}}$ to remove the menu bar from the graph. Now, press $\boxed{\text{EXIT}}$ and your calculator will display the previous graph, the one with the menu bar at the bottom. Press $\boxed{\text{EXIT}}$ again, and your calculator will return to the Home screen.

- Press $\boxed{\text{GRAPH}}$ to return to your graph and then press $\boxed{\text{CLEAR}}$. This time return directly to the Home screen by pressing $\boxed{\text{2nd}}$ $\boxed{\text{[QUIT]}}$.

That's it! You have completed the tutorial. Now practice and experiment on your own with the calculator until you begin to feel comfortable with these basic operations. The remainder of this guide will introduce new techniques as they are needed, chapter by chapter, for your work in *Precalculus: Concepts in Context*.

Chapter 1

General Information

Clearing Stored Functions

Before you get started on a problem, you probably will want to erase any functions that are stored in your calculator's memory. The tutorial demonstrated one method for clearing functions. Here is another.

- Press [2nd] [[MEM]] [F2] to access the MEM/DELETE menu shown below.

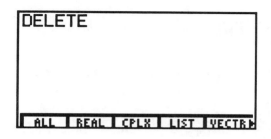

Note: If you press [F1] for ALL, you will erase <u>everything</u> that is stored in your calculator's memory, including programs, data, and functions. The MEM/DELETE menu allows you to be selective in what you choose to erase. In this case, we want to erase only the stored functions. These will be listed as equations under EQU.

- Press [MORE] to view other items in this menu. EQU should appear above the [F3] key. Press [F3] to select EQU. A list of all stored functions will appear. For example, the following screen indicates that three functions, $y1$, $y2$, and $y3$, are stored in memory.

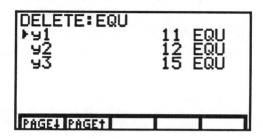

- Press [ENTER] repeatedly (once for each function stored) until all functions have been deleted.

TI-85 Guide

Making a Table of Values from a Formula

Example: Use the TI-85 to complete the following table for the function $y = 2x - 30$.

x	y
2.0	
6.0	
10.0	
14.0	
18.0	
22.0	
26.0	

Step 1: Enter the function $y = 2x - 30$.

- Press $\boxed{\text{GRAPH}}$ $\boxed{\text{F1}}$. Then enter $2x - 30$ as $y1$.

- Press $\boxed{\text{2nd}}$ $\boxed{\text{[QUIT]}}$ to return to the home screen.

Step 2: Compute the y-values corresponding to the x-values in the preceding table.

Plan of Action: We will use the TI-85's sequence command to generate the y-values corresponding to the x-values in the table. The sequence command needs the following information: the dependent variable ($y1$), the independent variable (x), the independent variable's starting value (2.0) and ending value (26.0), and the size of increment for the independent variable (4.0). The instructions below explain how to enter the command: seq($y1, x, 2, 26, 4$).

- Press $\boxed{\text{2nd}}$ $\boxed{\text{[CATALOG]}}$ for a listing of commands for the TI-85. A marker (black triangle) will appear opposite the command **abs**. Press $\boxed{\text{F1}}$ and/or $\boxed{\downarrow}$ to position the marker opposite the command **seq(** and then press $\boxed{\text{ENTER}}$.

- Press $\boxed{\text{2nd}}$ $\boxed{\text{[VARS]}}$ $\boxed{\text{MORE}}$ $\boxed{\text{F3}}$ for EQU. Position the marker opposite $y1$ and press $\boxed{\text{ENTER}}$.

- Complete the command by pressing $\boxed{,}$ $\boxed{x\text{-VAR}}$ $\boxed{,}$ $\boxed{2}$ $\boxed{,}$ $\boxed{2}$ $\boxed{6}$ $\boxed{,}$ $\boxed{4}$ $\boxed{)}$.

- To execute this command, press $\boxed{\text{ENTER}}$. Your screen should match the one that follows.

Chapter 1

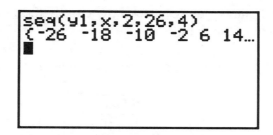

The sequence of y-values is listed horizontally. (-26 is the y-value associated with the x-value of 2, -18 is the y-value associated with the x-value of 6, and so forth.) Press the left arrow key $\boxed{\rightarrow}$ to see the y-values that go off the screen. Now, fill in the table entries for the y-column.

Notice that the dependent variable switches sign from negative to positive at some x-value between 14 and 18. In the next example, we'll pin-point the x-value corresponding to the switch in sign for the y-values.

Example: For $y = 2x - 30$, determine the x-value that corresponds to a zero y-value.

To solve this problem, we'll edit the sequence command used in the previous example so that it generates y-values corresponding to x-values between 14 and 18 separated by increments of 1.0.

- Press $\boxed{\text{2nd}}$ $\boxed{\text{[ENTRY]}}$ to return to the sequence command.

- Change the 2 to 14: position the cursor over 2 and press $\boxed{1}$ $\boxed{\text{2nd}}$ $\boxed{\text{INS}}$ $\boxed{4}$. Then change the 26 to 18: position the cursor over the 2 in 26 and press $\boxed{1}$ $\boxed{8}$. Change the 4 to 1: position the cursor over the 4 and press $\boxed{1}$. Finally, press $\boxed{\text{ENTER}}$.

You should now be able to determine the x-value that corresponds to a y-value of zero. (Did you get an x-value of 15?)

Lab 1: Fahrenheit

Plotting Points

You can use your calculator to plot the Fahrenheit-Celsius data given in the preparation for Lab 1. Then, graph your guess for the formula that relates degrees Fahrenheit to

TI-85 Guide

degrees Celsius. This will allow you to check how closely the function specified by your formula follows the pattern of the data.

Example: Plot the data in the following table and then overlay the graph of $y = 18x + 85$.

Sample Data	
x	y
-2	40
-1	60
1	100
3	140

Step 1: Adjust the viewing window as follows.

- Press $\boxed{\text{GRAPH}}$ $\boxed{\text{F2}}$.

- Choose a value for xMin that is smaller than the x-coordinates in the table and a value for xMax that is larger than the x-coordinates.

- Similarly, select appropriate settings for yMin and yMax.

- Decide on the spacing of the tick marks and set xScl and yScl. (What would be the disadvantage of setting yScl=1? How many tick marks would appear between 40 and 140?)

Step 2: Clear or turn off any stored functions. (See instructions on page 75.)

Step 3: Check to see if any data has been stored in lists named X or Y.

- Press $\boxed{\text{2nd}}$ $\boxed{\text{[MEM]}}$ $\boxed{\text{F2}}$ $\boxed{\text{F4}}$. The names of all lists containing data, in addition to xStat and yStat, will appear on your screen.

- If X and Y do not appear among the names in this list, proceed to the next step. Otherwise, follow the instructions in Step 7 to erase the data in lists X and Y.

- Press $\boxed{\text{EXIT}}$ or $\boxed{\text{2nd}}$ $\boxed{\text{[QUIT]}}$ to return to the home screen.

Step 4: Enter the sample data.

- Press $\boxed{\text{STAT}}$ $\boxed{\text{F2}}$ for EDIT. You will get a viewing screen similar to the one that follows.

Chapter 1

The cursor, a box with an A inside, will be blinking over the x in xStat. This style of cursor indicates that the ALPHA key is locked; therefore, when you press a key, you will get the upper blue key function.

- Type in X by pressing $\boxed{+}$, then press $\boxed{\text{ENTER}}$. Next, type in Y by pressing $\boxed{0}$ $\boxed{\text{ENTER}}$. Your screen should look like this:

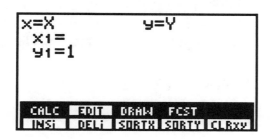

- Enter the data <u>one</u> x-y pair at a time. (The cursor will be blinking oppoiste $x1$.) Press $\boxed{\text{(-)}}$ $\boxed{2}$ $\boxed{\text{ENTER}}$ $\boxed{4}$ $\boxed{0}$ $\boxed{\text{ENTER}}$ for the first ordered pair. Then press $\boxed{\text{(-)}}$ $\boxed{1}$ $\boxed{\text{ENTER}}$ $\boxed{6}$ $\boxed{0}$ $\boxed{\text{ENTER}}$ for the second; press $\boxed{1}$ $\boxed{\text{ENTER}}$ $\boxed{1}$ $\boxed{0}$ $\boxed{0}$ $\boxed{\text{ENTER}}$ for the third; and press $\boxed{3}$ $\boxed{\text{ENTER}}$ $\boxed{1}$ $\boxed{4}$ $\boxed{0}$ $\boxed{\text{ENTER}}$ for the last ordered pair.

Step 5: Plot the sample data.

- Press $\boxed{\text{EXIT}}$ to return to the STAT menu.

- Press $\boxed{\text{F3}}$ to select DRAW and then $\boxed{\text{F2}}$ for SCAT (scatter plot).

- Press $\boxed{\text{CLEAR}}$ to clear the menu from the scatter plot. (If you have chosen appropriate settings for xMin, xMax, yMin, and yMax, you should see a plot of the four data points on your screen.)

Step 6: Plot the points from the table and display the graph of $y = 18x + 85$ in the same viewing window.

- In the Graph/$y(x) =$ menu, enter $18x + 85$ opposite $y1=$. (We assume that you have erased all other functions.) Press $\boxed{\text{EXIT}}$ $\boxed{\text{F5}}$ to graph this line. Notice that the scatter plot has disappeared from the screen.

- To overlay the scatter plot, press $\boxed{\text{STAT}}$ $\boxed{\text{F3}}$ $\boxed{\text{F2}}$ $\boxed{\text{CLEAR}}$. The plotted points should lie close to the line.

Step 7: Clear the lists named X and Y.

- Press $\boxed{\text{2nd}}$ $\boxed{\text{[MEM]}}$ $\boxed{\text{F2}}$ $\boxed{\text{F4}}$.

- Use the up $\boxed{\uparrow}$ or down $\boxed{\downarrow}$ arrow keys to position the triangle marker opposite X and then press $\boxed{\text{ENTER}}$.

- Next, position the marker opposite Y and press $\boxed{\text{ENTER}}$.

- Return to the Home screen by pressing $\boxed{\text{EXIT}}$.

Project 1.1: The Graphing Game

Adjusting the Viewing Window for Square Scaling

The viewing screen on your calculator is a rectangle. Therefore, if you use the standard window, the tick marks on the y-axis will be closer together than those on the x-axis. For square scaling we want the distance between 0 and 1 on the x-axis to be the same as the distance between 0 and 1 on the y-axis.

First, we'll observe the graph of the line $y = x$ in the standard viewing window and then we'll switch to square scaling.

Step 1: Graph $y = x$ in the standard viewing window.

- Erase any previously stored functions. (See the instructions on page 75.)

- Press $\boxed{\text{GRAPH}}$ $\boxed{\text{F1}}$ to access the GRAPH/$y(x)=$ menu

- Enter x opposite $y1$: press $\boxed{\text{F1}}$ for x, and then $\boxed{\text{EXIT}}$ to return to the $\boxed{\text{GRAPH}}$ menu.

- Press $\boxed{F3}$ to access the GRAPH/ZOOM menu, and $\boxed{F4}$ for ZSTD.

- Press $\boxed{CLEAR}$ to remove the menu from the bottom of the viewing screen.

Observe the spacing between the tick marks on the x and y-axes. Notice that the tick marks on the y-axis are closer together than the tick marks on the x-axis.

Step 2: Change to a square viewing window:

- Press $\boxed{GRAPH}$ $\boxed{F3}$ to access the GRAPH/ZOOM menu, and then press $\boxed{MORE}$ $\boxed{F2}$ for ZSQR. When the graph appears on your screen, observe the equal distance between tick marks on the two axes.

- Press $\boxed{EXIT}$ $\boxed{F2}$ and note the changes for xMin and xMax. (Recall that in the standard viewing window xMin $= -10$ and xMax $= 10$.)

- Press $\boxed{F5}$ $\boxed{CLEAR}$ to return to the graph of $y = x$.

The line $y = x$ should now appear to be inclined at a 45° angle, cutting the 90° angle made by the intersection of the x- and y-axes in half. The distances between the tick marks on the x- and y-axes should be the same.

TI-85 Guide

Chapter 2

General Information

Graphing a Quadratic Function

When you graph a quadratic function, it is important to experiment with various viewing windows to ensure that you have captured all the important features of the graph on your screen.

Example: Graph $y = 2x^2 - 2x + 12$ using the three different viewing windows specified in the directions below.

Start by entering the function:

- Clear all previously entered functions from your calculator's memory. (See instructions on page 75.)

- Press $\boxed{\text{GRAPH}}$ $\boxed{\text{F1}}$ and enter $2x^2 - 2x + 12$. (Remember to use the black subtraction key when entering $-2x$.)

- Press $\boxed{\text{EXIT}}$ to return to the GRAPH menu. Then press $\boxed{\text{F2}}$ for the RANGE settings.

Window 1: Adjust the settings to match those shown below and then press $\boxed{\text{F5}}$.

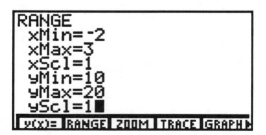

Your graph should be the familiar ∪-shape of a parabola.

Next, we view the graph of this quadratic function in two other windows. In each case, the viewing window selected fails to show key features of the parabola.

Chapter 2

Window 2: Change the xMin setting to 2 as follows.

- In the GRAPH menu, press $\boxed{\text{RANGE}}$.

- Press $\boxed{2}$ to change xMin to 2 and then press $\boxed{\text{F5}}$ to graph the function.

Notice in Window 2 the graph of $y = 2x^2 - 2x + 12$ looks more like a line than a parabola.

Window 3: Change to the standard viewing window as follows.

- While still in the graph menu, press $\boxed{\text{F3}}$ to access the GRAPH/ZOOM menu and then press $\boxed{\text{F4}}$ for ZSTD.

- Press $\boxed{\text{CLEAR}}$ to remove the menu bar from the graphing screen.

What do you see? In general, if you ask the calculator to graph a function, and just see empty axes, you probably are looking at a part of the plane that contains none of the graph.

Calculating Outputs of a Function

Example: Find the output of $f(x) = x^2 - x + 6$ when x has value 4.

- Press $\boxed{\text{GRAPH}}$ $\boxed{\text{F1}}$ and enter $x^2 - x + 6$ as $y1$.

- Press $\boxed{\text{EXIT}}$ $\boxed{\text{MORE}}$ $\boxed{\text{MORE}}$ $\boxed{\text{F1}}$ for EVAL.

- Enter the x value: press $\boxed{4}$. Then press $\boxed{\text{ENTER}}$.

The corresponding y-value of $18(= f(4))$ will appear at the bottom of your screen.

Programming the Quadratic Formula (Optional)

In order to solve inequalities and equations algebraically in Laboratory 2B and in some of the Chapter 2 exercises, you will need to use the quadratic formula. The steps involved in applying the quadratic formula can be stored as a program in your calculator.

TI-85 Guide

The program QUADFORM, designed to calculate the discriminant and solutions to a quadratic equation of the form

$$A x^2 + B x + C = 0.$$

is given in the table below. Instructions on entering and running this program follow. (We have placed a line number to the right of each command. Do not enter these numbers as part of your program.)

Quadratic Formula Program

Command	Line
NAME= QUADFORM	0
:Disp "INPUT A,B,C"	1
:Input A	2
:Input B	3
:Input C	4
:$B^2 - 4AC \to D$	5
:If D < 0	6
:Goto G	7
:$(-B - \sqrt{D})/(2A) \to E$	8
:Disp "X="	9
:Disp E	10
:$(-B + \sqrt{D})/(2A) \to F$	11
:Disp "X="	12
:Disp F	13
:Lbl G	14
:Disp "D="	15
:Disp D	16

Step 1: Preliminary Advice.

- If you want to leave the program at any time, press [2nd] [QUIT]. To return to the program you are writing, press [PRGM] [F2], for EDIT. Then press the F-key that corresponds to QUADFORM followed by [ENTER].

- If you need to insert a blank line below another line, move the cursor to the end of the line and press [2nd] [INS] [ENTER]. You can delete a line with [DEL].

- If you are multiplying variables with different names, you must use the multiplication key. For example, $A \cdot C$ must be entered by the key strokes

84

Chapter 2

ALPHA [A] × ALPHA [C]. A constant times a variable, such as 4A, does not require insertion of the multiplication sign; enter 4A by pressing 4 ALPHA [A].

- For additional help on programming, turn to the section on programming in your TI-85 manual.

Warning: Do not reset your calculator after entering your program or you will lose your program. Specifically, do not select ALL or PRGM in the MEM/DELET menu; do not select ALL or MEM in the MEM/RESET menu.

Step 2: Entering the Program.

After you have entered a command line, press ENTER to move to the next command line. The TI-85 will automatically insert a colon at the start of the new command line. Instructions relevant to individual command lines in QUADFORM follow.

To begin, press PRGM F2 for EDIT. Your calculator will prompt you for the name of the program.

Line 0: On your screen you should see **Name=** followed by a blinking cursor with an A inside. (This style cursor indicates that the ALPHA key has been activated.) Type in the name of the program QUADFORM. (The letters appear in alphabetical order on your calculator with [A] the blue upper function of the LOG key, and [Q] the upper function of the 4 key.) Remember that you must press ENTER to move to the next command line.

Lines 1 − 4. When you run this program, you will need to input the values of the coefficients of the quadratic function, $Ax^2 + Bx + C$. Create the input request prompts as follows:

- Press F3, to access the PGRM/I/O (input/output) menu, and then press F3 again for **Disp**(lay).

- To enter the remainder of command line 1:

  Press MORE F5 for ".

Press [ALPHA] [I] [ALPHA] [N] [ALPHA] [P] [ALPHA] [U] [ALPHA] [T] [ALPHA] [␣] (same key as [(-)]).

Press [ALPHA] [A] , [ALPHA] [B] , [ALPHA] [C].

Finally, press [F5] for ", and then [ENTER] to move to the next command line.

- To create lines 2 − 4, you will need to access the **Input** command. **Input** is the first selection in the PGRM/I/O menu. After entering the last set of quotation marks for line 1, you will need to press [MORE] to go back to the beginning of the options in the PGRM/I/O menu. Then press [F1] to select **Input**.

Line 5. The value of the discriminant is computed. To enter the formula for the discriminant:

- Press [ALPHA] [B] x^2 [−] [4] [ALPHA] [A] [×] [ALPHA] [C].

- Then press [STO▷] for store as, followed by [D]. (Note that pressing [STO▷] activates the [ALPHA] key.)

Here's what lines 6 and 7 do. If the discriminant is negative, there are no real solutions to the quadratic equation. In this case, no further calculations will be made and only the discriminant will be displayed. This is accomplished as follows. When D is negative, the **If** statement in line 6 is true and the program executes the **Goto** command in line 7. Executing the **Goto** command causes the program to jump down to **Lbl G** (line 14) thus bypassing any further calculations.

Enter line 6 as follows.

- To key in **If**, you will first need to press [EXIT] to exit the PGRM/I/O menu and return to the PRGM menu. Your screen should match the one below.

86

Chapter 2

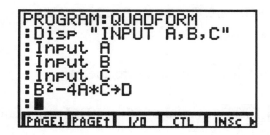

Press F4 to open the PRGM/CTL menu. Then press F1 to insert the **If** command.

- Press ALPHA [D].

- To insert the inequality sign, <, press 2nd [TEST] F2. Press EXIT to return to the previous menu.

- Press 0 to complete the command in Line 6.

Now enter Line 7 as follows.

- **Goto** is in the PGRM/CLT menu. After entering line 6, you should be in this menu. Press MORE F5 for **Goto**.

- Press ALPHA [G]

Lines 8 − 13: If the if-statement in Line 6 is false, the program jumps to line 8. This happens when the discriminant is positive or zero and there are two real solutions. (Note the two solutions are the same when D = 0.) Lines 8 − 13 compute and display the two solutions.

- In line 8, remember to use the (-) key for the opposite of B and the subtraction key − between B and $\sqrt{D}$.

- In lines 9, 12, and 15, press ALPHA [=] (same key as STO▷) to get the equals sign.

Lines 14 − 16. **Lbl G** marks the place where the program is resumed if the discriminant is negative. Lines 15 and 16 display the value of the discriminant.

- The command **Lbl** is in the PRGM/CTL menu. You'll need to press EXIT to return to the PGRM menu from the PGRM/I/O menu and then press F4

87

TI-85 Guide

to enter the PGRM/CTL menu. Press MORE to see more options. Press F4 to insert **Lbl**.

- Press ALPHA [G] to complete line 14.

- Next, press EXIT to return to the PRGM menu and then F3 to enter the PRGM/I/O menu.

- Press F3 to insert **Disp**. Finish entering lines 15 and 16.

Once you have entered the entire program, press 2nd [QUIT] to return to the Home screen.

Step 3: Debugging the Program.

This is best done by running the program and letting the TI-85 identify the errors. Then, after you get the program to run, take an equation and find the solutions by hand. Test the program using this equation so you can check that it is giving you the correct solutions.

(1) Let's try to run QUADFORM using the equation

$$2x^2 + 5x - 3 = 0.$$

If you encounter an error along the way, jump ahead to the instructions in (2).

- To run QUADFORM, press PRGM F1 for a list of the names of programs stored in your calculator. To execute the program, press the F-key that corresponds to QUADFORM and then press ENTER. If you have entered the first four lines of your program correctly, a prompt requesting the coefficients, A, B, and C, should appear:

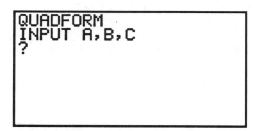

Chapter 2

- Enter the coefficients of the quadratic: press [2] [ENTER] [5] [ENTER] [(-)] [3] [ENTER]. If all goes well, your screen should match the one below, showing two roots, −3 and .5, and a discriminant of 49.

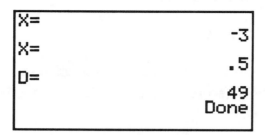

(2) Correct any errors that you have made in entering your program.

- If you have made errors that the calculator recognizes as errors, you will get an error message when you execute the program. For example, if in line 9 you used the subtraction key for -B, the following error would appear when you tried to execute the program.

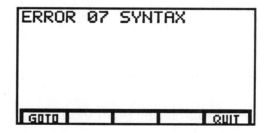

If you press [F1] for GOTO, you will return to the program and the cursor will mark the location of the first error that your TI-85 encountered in executing the program. Correct the error and try running the program again. Repeat this process until all errors have been corrected.

- After you get the program to run, check that you get the correct solutions. If your solutions do not agree with the ones given on the preceding page, check that you have entered lines 5, 8, and 11 correctly.

Step 4: Transferring the program from one calculator to another.

After entering a lengthy program, you may want to transfer your program to another student's calculator. You will need the your calculator's link cable in order to do this.

TI-85 Guide

- Connect the calculator with the program to the calculator without the program using the link cable. (There is a small hole for this cable at the bottom of your calculator.) Make sure that the cables are pushed all the way in to form a good connection.

- Turn both calculators on.

- The person who wants to receive the program should press $\boxed{\text{2nd}}$ $\boxed{\text{LINK}}$ $\boxed{\text{F2}}$ for RECV (short for receive). The message **Waiting** will then appear on their screen.

- Next, the person who wants to send the program should press $\boxed{\text{2nd}}$ $\boxed{\text{LINK}}$ $\boxed{\text{F1}}$. You are then given a number of choices of types of information to send. Press $\boxed{\text{F2}}$ for PRGM. If necessary, position the black marker opposite QUADFORM. Then press $\boxed{\text{F1}}$ for XMIT (short for transmit). When the transmission is complete, the message **Done** will appear on your screen.

Laboratory 2A: Galileo

Approximating the Coordinates of a Point on a Graph.

Example: Approximate the vertex (turning point) of the parabola $y = 2x^2 - 2x + 12$.

Step 1: Graph $y = 2x^2 - 2x + 12$ and get a rough approximation for the vertex.

- Press $\boxed{\text{GRAPH}}$ $\boxed{\text{F1}}$ to access the GRAPH/$y(x) =$ menu. Enter the function $2x^2 - 2x + 12$.

- Press $\boxed{\text{EXIT}}$ $\boxed{\text{F2}}$ and adjust the RANGE settings to match the ones on the screen below.

```
RANGE
 xMin=-3
 xMax=3
 xScl=1
 yMin=5
 yMax=15
 yScl=1
 y(x)= RANGE ZOOM TRACE GRAPH▸
```

Chapter 2

- Press $\boxed{\text{F4}}$ for TRACE. Use the right $\boxed{\rightarrow}$ and left $\boxed{\leftarrow}$ arrow keys to position the cursor (a box with a blinking ×) at the vertex of the parabola. Read off the estimates for the x- and y-coordinates from the bottom of the screen.

Step 2: Next, let's improve the accuracy of the estimate by magnifying the portion of the graph near the vertex of the parabola.

- Press $\boxed{\text{GRAPH}}$ $\boxed{\text{F3}}$ for ZOOM and then press $\boxed{\text{F1}}$ for Box. A free floating cursor (four dots surrounding a blinking center dot) will appear in the center of your screen. Press the left arrow key $\boxed{\leftarrow}$ to move the cursor off the y-axis so that you can see it clearly.

- Draw a box surrounding the vertex of the parabola as follows. Use the arrow keys to position the cursor at the upper left-hand corner of your box. (Refer to the picture below.) Press $\boxed{\text{ENTER}}$. Now, press the right $\boxed{\rightarrow}$ and down $\boxed{\downarrow}$ arrow keys to move the lower right-hand corner of the box. When the picture on your screen resembles the one that follows, press $\boxed{\text{ENTER}}$. (The coordinates of the lower right corner of the box appear at the bottom of the screen. The coordinates on your screen will probably differ from those below.)

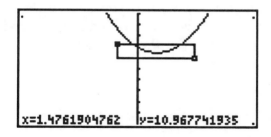

- Press $\boxed{\text{EXIT}}$ to return to the GRAPH menu and then $\boxed{\text{F4}}$ for TRACE. Position the cursor at the vertex to approximate its coordinates. (The exact coordinates of the vertex are (.5, 11.5). After zooming in, your estimate for the vertex should be close to the exact.)

Notice that the box drawn about the vertex of the preceding parabola is short and wide. When we zoom in with a short-wide box, the sides of the parabola near the vertex become quite steep. If we had magnified a section of the parabola near the vertex using a square-shaped box or a tall-narrow box, the magnified section would have been flatter. It is difficult to pin-point the vertex when the section of graph surrounding the vertex is fairly flat.

TI-85 Guide

Laboratory 2B: Electric Power

Finding the Points of Intersection of Graphs

Example: Find the points where the graphs of $f(x) = -2x^2 + 2x + 12$ and $g(x) = -3x + 14$ intersect.

You could approximate the solutions to this problem using the techniques outlined in the example on page 71. However, the TI-85's ISECT command will provide exact solutions when the coordinates of the point of intersection have finite decimal representations that are short enough to fit on the display at the bottom of the screen. In other cases, you will get a very good approximation without having to zoom in on a section of the graph.

Step 1: Find a viewing window that gives a clear view of the points of intersection.

- Enter the two functions into your calculator.

- Adjust the RANGE settings so that you have a clear view of both points of intersection. (Hint: You might start with the standard viewing window, STD in the GRAPH/ZOOM menu, and then adjust the RANGE settings after viewing the graph.)

Step 2: Approximate the coordinates of one of the points of intersection as follows.

- From the GRAPH menu press MORE and then F1 to access the GRAPH/MATH menu. Then press MORE again. Your screen now look like this:

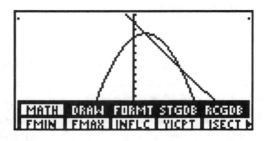

- Press F5 to select ISECT (short for intersection). Your screen should resemble the one that follows.

92

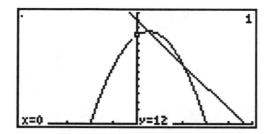

Notice that a cursor is positioned on one of the curves (on the screen above the cursor is at the point (0, 12) on the parabola) and the number 1 appears in the upper right corner of your screen. (The calculator needs to know which two curves you want to work with. In this particular case there are only two curves, so if you press the up ↑ or down ↓ arrow keys the cursor will jump back and forth from the line to the parabola.)

- Position the cursor on either the line or the parabola and press ENTER. The cursor will automatically jump to the other curve. (<u>Don't</u> press ENTER yet.)

- Now you must provide your calculator with a guess for the point of intersection. Use the right → or left ← arrow keys to move the cursor to one of the points of intersection and press ENTER. After a brief wait, you'll see the coordinates of this point of intersection displayed at the bottom of your screen.

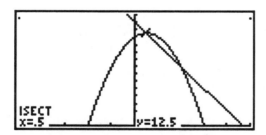

Step 3: Find the coordinates of the second point of intersection.

- Press GRAPH.

- Determine the coordinates of the second point of intersection by repeating the process outlined in Step 2.

If you have done everything correctly, you will find that the two graphs intersect at (.5, 12.5) and (2, 8).

TI-85 Guide

Graphing Piecewise-Defined Functions

Example: Graph the piecewise-defined function $f(x) = \begin{cases} x - 4 & \text{if } x > 4 \\ -x + 4 & \text{if } x \leq 4 \end{cases}$.

The graph of $f(x)$ consists of two half-lines pieced together. You'll want the graph of $y = x - 4$ when x-values are greater than 4 and $y = -x + 4$ when x-values are less than or equal to 4.

Step 1: Erase any previously stored functions. (If you have forgotten how to erase stored functions, turn to page 75.) Then from the GRAPH/ZOOM menu, press [F4] to set the standard viewing window.

Step 2: Enter the functions that you want to piece together.

- From the GRAPH menu, press [F1] to enter the GRAPH/$y(x)=$ menu.

- Enter $y = x - 4$ as $y1$ and $y = -x + 4$ as $y2$.

- Press [EXIT] to return to the GRAPH menu. Then press [F5] to graph the two functions. Your graph should look like an $\times$.

Step 3: Form the function $f(x)$. This is where you'll piece together the graphs of $y1$ and $y2$.

- Press [F1] and then press [ENTER] twice to position the cursor opposite $y3=$.

- Enter $f(x)$ as $y3$: press [F2] [1] to insert $y1$.

- Press [×]. Next, enter the condition that governs when to use $y1$: press [( ] [F1] [2nd] [TEST] [F3] [4] [ )]. Then press [EXIT] to return to the GRAPH/$y(x)=$ menu.

- Press [+]

- Enter function $y2$: press [F2] [2] to insert $y2$.

- Press [×]. Then enter the condition that governs when to use $y2$: press [( ] [F1] [2nd] [TEST] [F4] [4] [ )]. When you have completed this step your screen should match the one that follows.

94

Chapter 2

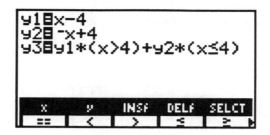

Here's how your calculator interprets the information you've just entered as a piecewise-defined function. The calculator assigns the expression $(x > 4)$ the value 1 when the inequality is true (in other words, when the input variable, x, is greater than 4); when the inequality is false, the calculator sets the expression $x \leq 4$ equal to 0. Thus, for $x > 4$, the function $y3$ is equivalent to:

$$y3 = (x-4)(1) + (-x+4)(0) = x - 4.$$

And when $x \leq 4$ the function $y3$ is equivalent to:

$$y3 = (x-4)(0) + (-x+4)(1) = -x + 4.$$

Step 4: Graph $f(x)$.

- Press $\boxed{\text{EXIT}}$ to return to the GRAPH/$y(x) =$ menu.

- Turn off $y1$ and $y2$ as follows. Position the cursor opposite $y1=$. Press $\boxed{\text{F5}}$ for SELCT (select) to remove the highlighting from $y1$'s equals sign. Then position the cursor opposite $y2=$ and press $\boxed{\text{F5}}$.

- Press $\boxed{\text{EXIT}}$ $\boxed{\text{F5}}$. The graph of $f(x)$ should look V-shaped.

Project 2.5: The Least Squares Line

Computing the Least Squares Line

If your data, when plotted, lies exactly on a line, you can use algebra to determine the equation of the line. However, real data seldom fall precisely on a line. Instead, the plotted data may exhibit a roughly linear pattern. The least squares line (also called the regression line) is a line that statisticians frequently use when describing a linear trend in data.

TI-85 Guide

Example: Use the least squares line to describe the linear pattern in the data below.

x	y
-3.0	-6.3
-2.0	-2.8
1.2	2.0
2.0	4.1
3.1	4.7
4.2	5.6

Step 1: Erase any stored functions. Then enter and the data from the table above into lists X and Y. (Refer to *Plotting Points* on pages 77 − 80.) After you have entered your data, press $\boxed{\text{2nd}}$ $\boxed{\text{[QUIT]}}$ to return to the home screen.

Step 2: Next, we'll find the equation for the least squares line. Statisticians specify the form of this line as follows: $y = a + bx$. Here b represents the slope of the line and a, the y-intercept.

- Press $\boxed{\text{STAT}}$ $\boxed{\text{F1}}$ for CALC.

- If $x$list and $y$list are not named X and Y as in the screen above, adjust their names by pressing $\boxed{\text{[X]}}$ $\boxed{\text{ENTER}}$ $\boxed{\text{[Y]}}$ $\boxed{\text{ENTER}}$. If $x$list and $y$list are already named X and Y, press $\boxed{\text{ENTER}}$ twice. Your screen should match the one below.

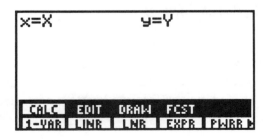

Chapter 2

- Press $\boxed{F2}$ for LINR. If you have entered your data correctly, your output will match this screen.

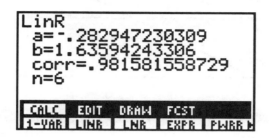

- Obtain the least squares line (regression line) for the data in the table by substituting the calculated values for a and b into the equation $y = a + bx$. Write your equation in the space provided below.

$$y = \underline{\hspace{1cm}} + \underline{\hspace{1cm}} x$$

Step 3. Graph the least squares line and the data in the same viewing window.

- Press $\boxed{\text{GRAPH}}$ $\boxed{F1}$. Then enter the function that you have written above; or press the following keys and the TI-85 will insert the regression equation for you: $\boxed{\text{2nd}}$ $\boxed{\text{[VARS]}}$ $\boxed{\text{MORE}}$ $\boxed{\text{MORE}}$ $\boxed{F3}$, use the down arrow key $\boxed{\downarrow}$ to move the marker opposite RegEq, and then press $\boxed{\text{ENTER}}$.

- Now press $\boxed{\text{EXIT}}$ $\boxed{F2}$ and adjust the RANGE settings so that all data points, when plotted, will appear on your screen.

- Press $\boxed{\text{STAT}}$ $\boxed{F3}$ for DRAW and a graph of the regression line will appear on your screen. It should look like this:

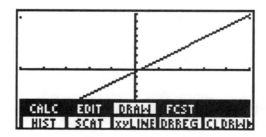

- Press $\boxed{F2}$ to add the scatter plot of the data points to the display. Then press $\boxed{\text{CLEAR}}$ to remove the menu from the bottom your screen. Your screen should look similar to that follows. (You may have chosen different RANGE settings.)

TI-85 Guide

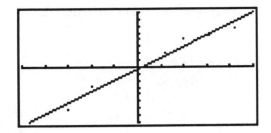

- To remove the scatter plot from your screen, press [STAT] [F3] [F5] for CLDRW (Clear Draw).

Chapter 3

Chapter 3

Lab 3: Graph-Trek

In Lab 3 you will be investigating the effect that certain algebraic modifications, such as adding a constant to the input variable, have on the graph of a function. You'll want to experiment using several different functions. We've provided some functions and algebraic modifications that you might want to consider.

Functions Involving Square Roots

Example: Compare the graphs of $y = \sqrt{x}$, $y = \sqrt{x+2}$, and $y = \sqrt{x} + 2$.

- Clear any previously stored functions. (Refer to the instructions on page 75.)

- Enter $y = \sqrt{x}$: Press $\boxed{\text{GRAPH}}$ $\boxed{\text{F1}}$ $\boxed{\text{2nd}}$ $\boxed{[\sqrt{\ }]}$ $\boxed{\text{F1}}$.

- Enter $y = \sqrt{x+2}$: Press $\boxed{\text{ENTER}}$ to position the cursor opposite $y2$. Then press $\boxed{\text{2nd}}$ $\boxed{[\sqrt{\ }]}$ $\boxed{(}$ $\boxed{\text{F1}}$ $\boxed{+}$ $\boxed{2}$ $\boxed{)}$.

- Finally, enter the function $y = \sqrt{x} + 2$: Press $\boxed{\text{2nd}}$ $\boxed{[\sqrt{\ }]}$ $\boxed{\text{F1}}$ $\boxed{+}$ $\boxed{2}$. (Notice that the parentheses are the only difference between the previous function and this one.)

- Press $\boxed{\text{EXIT}}$ $\boxed{\text{F5}}$ to view the graphs of the three functions that you have entered. (You may need to adjust the RANGE settings.)

Trigonometric Functions and the Trig Viewing Window

Locate the $\boxed{\text{SIN}}$, $\boxed{\text{COS}}$, and $\boxed{\text{TAN}}$ keys on your calculator. These keys will allow you to study the graphs of the sine, cosine, and tangent functions before the functions are formally introduced in Chapter 6. Prior to entering any of these functions, press $\boxed{\text{2nd}}$ $\boxed{\text{[MODE]}}$. If Radian is not highlighted, move the cursor to Radian and press $\boxed{\text{ENTER}}$.

TI-85 Guide

Example: Examine the graphs of $y = \sin(x)$ and $y = \sin(x + 2)$.

- Clear any previously stored functions, then press $\boxed{\text{GRAPH}}$ $\boxed{\text{F1}}$ for the GRAPH/$y(x) =$ menu.

- Enter the function $\sin(x)$ as $y1$: press $\boxed{\text{SIN}}$ $\boxed{\text{F1}}$.

- Next, enter the function $\sin(x + 2)$ as $y2$: press $\boxed{\text{SIN}}$ $\boxed{(}$ $\boxed{\text{F1}}$ $\boxed{+}$ $\boxed{2}$ $\boxed{)}$

- After pressing $\boxed{\text{EXIT}}$ to return to the GRAPH menu, press $\boxed{\text{F3}}$ $\boxed{\text{MORE}}$ $\boxed{\text{F3}}$ for ZTRIG and then $\boxed{\text{CLEAR}}$. ZTRIG will adjust the RANGE settings to the trigonometric viewing window settings. You should see two wavy curves. (If you don't, go back and check again that you are in Radian mode.)

- Finally, press $\boxed{\text{EXIT}}$ then $\boxed{\text{F2}}$ and observe RANGE settings for the trigonometric window.

Graphing a Family of Functions

Using your calculator's list capabilities, you can substitute each value in a given list for a constant in an algebraic formula. This feature allows you to graph an entire family of functions quickly. On the TI-85, you specify a list by enclosing the members of the list in brackets: { }.

Example: Graph the family of functions $y = (x + 1)^2$, $y = (x + 2)^2$, and $y = (x + 3)^2$.

- Clear any previously stored functions from your calculator's memory, then press $\boxed{\text{GRAPH}}$ $\boxed{\text{F2}}$ and change the RANGE settings to match those shown below.

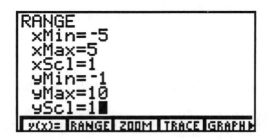

- After pressing $\boxed{\text{F1}}$ to access the GRAPH/$y(x) =$ menu, enter the three functions by specifying the constants, 1, 2, and 3, in a list as follows. Press $\boxed{(}$ $\boxed{\text{F1}}$ $\boxed{+}$ $\boxed{\text{2nd}}$ $\boxed{\text{[LIST]}}$ $\boxed{\text{F1}}$ for {, $\boxed{1}$, $\boxed{2}$, $\boxed{3}$ $\boxed{\text{F2}}$ for }, $\boxed{)}$ $\boxed{x^2}$. If you have entered the family of functions correctly, your screen will match the one that follows.

Chapter 3

- Press EXIT twice to return to the GRAPH menu, then press F5 . Watch as the three functions are graphed one after the other.

The Difference between $(-x)^2$ *and* $-x^2$

Example: Graph the functions $f(x) = (-x)^2$ and $g(x) = -x^2$ in the standard viewing window.

- Clear any previously stored functions, then press GRAPH   F3   F4 to set up the standard viewing window.

- After pressing F1 to access the GRAPH/$y(x) =$ menu, enter the function $f(x) = (-x)^2$: press ((-)   F1   ) x^2 .

- Now enter the function $f(x) = -x^2$, press (-)   F1 x^2 .

- Press EXIT   F5 to view the graph.

Which of these functions has negative outputs?

Projects 3.1 and 3.2: Vertical and Horizontal Stretching and Compression

This section assumes that you are familiar with the TI-85's list capabilities. You may want to review *Graphing a Family of Functions* on page 100 before working through this section.

TI-85 Guide

Entering the Family of Functions, c · f(x)

Example: To graph the family of functions $y = x^2$, $y = 2x^2$, and $y = .5x^2$, enter the family as $\{1, 2, .5\}x^2$.

Entering the Family of Functions, f(c · x)

Example: To graph the family $y = x^2$, $y = (2x)^2$, and $y = (.5x)^2$, enter the family as $(\{1, 2, .5\}x)^2$.

Project 3.6: Absolute Value in Functions

Functions Involving Absolute Value

Example: Graph $y = |x|$ and $y = |x + 2|$ in the standard viewing window.

- Clear any previously stored functions. (See page 75.)

- Set up the standard viewing window by pressing ⎡GRAPH⎤ ⎡F3⎤ ⎡F4⎤. Then press ⎡F1⎤ for the GRAPH/$y(x) =$ menu.

- To enter $y = |x|$ as $y1$, press ⎡F1⎤ for the GRAPH/$y(x) =$ menu, then press ⎡2nd⎤ ⎡[CATALOG]⎤ ⎡ENTER⎤ for **abs** (short for the absolute value), ⎡F1⎤.

- To enter $y = |x + 2|$ as $y2$: press ⎡2nd⎤ ⎡[CATALOG]⎤ ⎡ENTER⎤ ⎡(⎤ ⎡F1⎤ ⎡+⎤ ⎡2⎤ ⎡)⎤.

- Press ⎡EXIT⎤ ⎡F5⎤ to graph the two functions.

Note: *When you want the absolute value of an expression, you must enclose the entire expression in parentheses.*

Chapter 4

Chapter 4

General Information

Determining the Roots of a Function

Determining the roots (zeros, or x-intercepts) of a function using factoring often requires considerable skill; and most polynomials can't be factored easily, if at all. However, your calculator can take much of the drudgery out of finding roots.

Example: Find the roots of the polynomial function $g(x) = \frac{1}{2}x^4 + 2x^3 + x^2 - 3x - 5$.

Step 1: Graph $g(x)$ in the standard viewing window.

- Clear any previously stored functions.

- Enter $g(x)$ as $y1$. (Note: The entire function will not fit on your screen. Instead of the first term of the polynomial, you will see ellipsis marks (...). If you want to view the first term, move the cursor to the front of the polynomial by pressing $\boxed{\text{2nd}}$ $\boxed{\leftarrow}$.)

- Select the standard viewing window and graph the function. (If you need some help, refer to Topic 6, *Graphing* (page 69).

Step 2: Let's approximate the negative root first. (You could zoom in on this intercept and then use TRACE to estimate the x-coordinate. However, here is a way to determine the root without zooming in.)

- From the GRAPH menu, press $\boxed{\text{MORE}}$ $\boxed{\text{F1}}$ to access the GRAPH/MATH menu.

- To approximate the negative root, you'll have to specify a narrow interval about the negative x-intercept. Press $\boxed{\text{F1}}$ for LOWER. Using the left arrow key $\boxed{\leftarrow}$, move the cursor slightly to the left of the negative x-intercept and then press $\boxed{\text{ENTER}}$ to mark with a black triangle the lower bound of the interval.

- Next, press $\boxed{\text{F2}}$ for UPPER. Use the right arrow key $\boxed{\rightarrow}$ to position the cursor slightly to the right of the negative root (but still to the left of the positive root) and press $\boxed{\text{ENTER}}$ this time marking the upper bound of the interval.

TI-85 Guide

- Now, press F3, for ROOT. You'll have to provide the calculator with a guess for the root: use ← and/or → to move the cursor as close to the negative root as possible and then press ENTER.

At this point, your screen should look similar to the one shown below. The black triangles mark an interval that contains the root, the cursor marks the root on the graph, and the line at the bottom of the screen displays the approximate coordinates of the root.

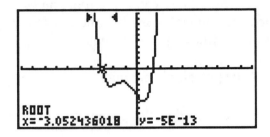

Step 3: To approximate the positive root, press GRAPH and then adapt the instructions in Step 2. The marker triangles will move when you select new upper and lower bounds. (They will disappear from your screen if you alter any of the RANGE settings.) After adapting the instructions, you should get approximately 1.39 for this root.

Lab 4A: Packages

Finding Local Maxima or Minima of Functions

In Lab 2A, you found the vertex of a parabola using a procedure requiring a combination of ZOOM/BOX and TRACE. (See *Approximating the Coordinates of a Point on a Graph* pages 90 and 91 for details.) The coordinates of any function's turning points could be estimated using this method. One drawback to this procedure is that you must frequently apply it several times in succession before you can obtain the desired accuracy. Here is another method for finding local maxima and/or minima of a function.

Example: Let's estimate the local maximum and local minimum of the cubic function

$$f(x) = x^3 - 4x^2 + 2x - 4.$$

Step 1: Graph $f(x)$ using a viewing window that gives you a clear view of the two turning points (one peak and one valley) of the graph.

Chapter 4

Step 2: Approximate the coordinates of the turning point associated with the local maximum (the y-coordinate of the peak on the graph) as follows.

- Press $\boxed{\text{MORE}}$ $\boxed{\text{F1}}$ for the GRAPH/MATH menu. Use LOWER and UPPER to specify an x-interval that contains the x-coordinate associated with the local maximum of the function. (Refer to Step 2 on page 103 for help in setting up this interval.)

- While still in the GRAPH/MATH menu, press $\boxed{\text{MORE}}$ $\boxed{\text{F2}}$ for FMAX. Your calculator needs a guess for the turning point. So, you'll need to use $\boxed{\leftarrow}$ and/or $\boxed{\rightarrow}$ to position the cursor on the turning point where the graph peaks and then press $\boxed{\text{ENTER}}$. Your screen should look similar to the one below which shows a local maximum value of approximately -3.73.

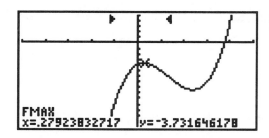

Step 3: Approximate the coordinates of the turning point associated with the local minimum (the y-coordinate of the valley on the graph).

- Press $\boxed{\text{GRAPH}}$ $\boxed{\text{MORE}}$ $\boxed{\text{F1}}$ to access the GRAPH/MATH menu. Use LOWER and UPPER to specify an x-interval that contains the x-coordinate associated with the local minimum of the function.

- Press $\boxed{\text{MORE}}$ $\boxed{\text{F1}}$ for FMIN. Use $\boxed{\leftarrow}$ and/or $\boxed{\rightarrow}$ to position the cursor on the turning point that resembles a valley. (This will give your calculator a guess for the coordinates of this turning point.) Then press $\boxed{\text{ENTER}}$. You should get a value for y that is close to -8.42.

Projects 4.1 and 4.2: Exploring Polynomial Graphs

Determining a Good Viewing Window for a Polynomial Function

It is frequently helpful to have some information about the range of a function's outputs prior to determining reasonable values for the window settings.

TI-85 Guide

Example: Graph $f(x) = 5x^3 - 9x^2 - 40x + 15$ in the standard viewing window.

After viewing the graph of $f(x)$ in the standard viewing window, you should realize that the settings for yMin and yMax need to be changed. But, by how much should you decrease yMin or increase yMax in order to get a more comprehensive view of this graph? One help is to evaluate the function at a couple of specific values.

- Approximate the outputs $f(-1)$ and $f(2)$. (If you've forgotten how to use EVAL to compute the output of a function refer to *Calculating Outputs of a Function*, page 83.) Did you get the value 41 for $f(-1)$?

- Adjust the settings on the viewing window so that yMin is somewhat smaller than $f(2)$ and yMax is somewhat larger than $f(-1)$.

The graph of $f(x)$ should now look much more like a classic cubic with two turning points.

Lab 4B: Doormats

Graphing a Rational Function

If the numerator or the denominator of a rational function consists of more than one term, you must enclose it in parentheses when you enter it into your calculator.

Example: Graph $r(x) = \dfrac{x^2 - 1}{x}$ using a standard viewing window.

- Press $\boxed{\text{Graph}}$ $\boxed{\text{F1}}$ and erase any previously stored functions. (Position the cursor on a function and press $\boxed{\text{CLEAR}}$.)

- Enter $r(x)$ as y1: press $\boxed{(}$ $\boxed{\text{F1}}$ $\boxed{x^2}$ $\boxed{-}$ $\boxed{1}$ $\boxed{)}$ $\boxed{\div}$ $\boxed{\text{F1}}$.

- Press $\boxed{\text{EXIT}}$ $\boxed{\text{F3}}$ $\boxed{\text{F4}}$ to view the graph in the standard window. (Draw a quick sketch of this graph so you remember what it looks like.)

Now, let's see what would happen if you forgot to enclose the numerator of $r(x)$ in parentheses.

- Press $\boxed{\text{F1}}$ to enter the GRAPH/$y(x) =$ menu.

Chapter 4

- Use DEL to delete the parentheses that surround the numerator.

- Press EXIT   F5 to view the graph. Did removing the parentheses affect the shape of the graph?

The formula for the function whose graph now appears on your screen is $x^2 - \frac{1}{x}$. Without parentheses, the calculator divides only the last term, 1, by x.

Zooming Out

Example: Graph the function $f(x) = \dfrac{x^2 - 1}{x - 3}$ in the standard viewing window.

Once more with feeling: When you enter a rational function, be sure to enclose both the numerator and denominator in parentheses; if you need help graphing the function in this example, refer to the instructions for the previous example. If you have entered the function correctly, your graph should match the one on the screen below.

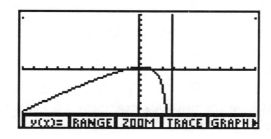

The vertical line in the graph above indicates that this function has a vertical asymptote at $x = 3$. Remember, this line is not part of the graph of the function. Furthermore, because the domain of this function is all real numbers except $x = 3$, there is a branch of this graph that lies to the right of the line $x = 3$. In order to observe this branch, you would have to adjust the RANGE setting for yMax. We'll do that on page 109.

Now, let's see what happens to the appearance of the graph of $f(x)$ as we "back away" by increasing the width and height of the viewing window.

- Press F2 to access the GRAPH/RANGE menu. Set xScl and yScl equal to 0. (This turns off the tick marks that appear on the axes. If you skip this step, the axis will get crowded with tick marks when you zoom out.)

- Check the settings for the zoom factor: press F3 to access the GRAPH/ZOOM menu, and then press MORE   MORE   F1 for ZFACT. Your screen should match the one below. If it doesn't, adjust the factor settings before proceeding to the next step.

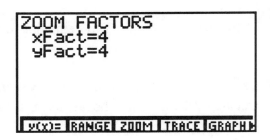

- Press $\boxed{F3}$ to re-enter the GRAPH/ZOOM menu, then press $\boxed{F3}$ a second time for ZOUT. (A blinking pixel, part of a free moving cursor, should appear in the center of your screen. The zooming will be centered around this location. If you wish to change the focal point of the zooming, use the arrow keys to position this cursor at a different center.)

- Press $\boxed{\text{ENTER}}$ to view the graph over wider x- and y-intervals. Press $\boxed{\text{ENTER}}$ again to zoom out a second time. (The graph should look like a line except, perhaps, for a small blip slightly to the right of the origin.)

- Now, press $\boxed{\text{GRAPH}}$ $\boxed{F2}$ to observe the affect on the RANGE settings of twice zooming out (by a factor of 4).

The default setting for ZOUT widens both the x- and y-intervals by a factor of 4 each time that it is applied. In the previous example, you zoomed out twice. Therefore, the x- and y-intervals are 16 times wider than they were before you zoomed out.

In the next example we'll change the default zoom settings in order to observe a graph of a function that begins to act like its horizontal asymptote.

Example: Graph $q(x) = \dfrac{5x^2 + 20x - 105}{2x^2 + 2x - 60}$ in the standard viewing window.

If you have entered the function correctly, your graph should look like this:

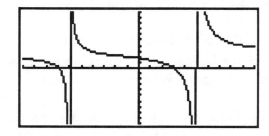

Chapter 4

Based on this graph you may suspect that the function $q(x)$ has a horizontal asymptote but it is not at all obvious. Let's observe the function's graph over increasingly wide x-intervals to see if it begins to behave like a horizontal line. Instructions on how to zoom out in the horizontal direction follow.

- From the GRAPH menu, press $\boxed{F2}$ and set xScl and yScl to 0.

- Press $\boxed{F3}$ to access the GRAPH/ZOOM menu. Press $\boxed{MORE}$ twice, followed by $\boxed{F2}$ for ZOOMX and then $\boxed{ENTER}$. (ZOOMX widens the x-interval by a factor of 4 and leaves y-interval unchanged.)

- Press $\boxed{ENTER}$ several more times to continue widening the x-interval. Your graph should begin to resemble its horizontal asymptote $y = 2.5$.

Project 4.4: Vertical Asymptotes and "Black Holes"

Vertical Asymptotes

Example: Let's return to the graph of $y = \dfrac{x^2 - 1}{x - 3}$ in the standard viewing window. (Refer to the calculator screen on page 107.)

For this example, your calculator draws the vertical asymptote at $x = 3$. However, this viewing window does not show any graph to the right of $x = 3$. Since the domain of this function includes all real numbers larger than 3, we need to adjust the viewing window in order to see what the graph looks like for x-values larger than 3.

Let's change the viewing window so that we can observe the shape of the branch of the graph that lies to the right of the vertical asymptote.

- Press $\boxed{F2}$ and change yMax to 20.

- Press $\boxed{F5}$ to view the graph.

Warning: *In the previous example, the TI-82 drew the vertical asymptote for you. However, in some viewing windows, the calculator does not draw the vertical asymptote. Therefore, you will need to keep track of the vertical asymptotes yourself and not rely exclusively on the calculator.*

109

TI-85 Guide

Let's change the viewing window to match the one below and look at one last graph of $f(x)$.

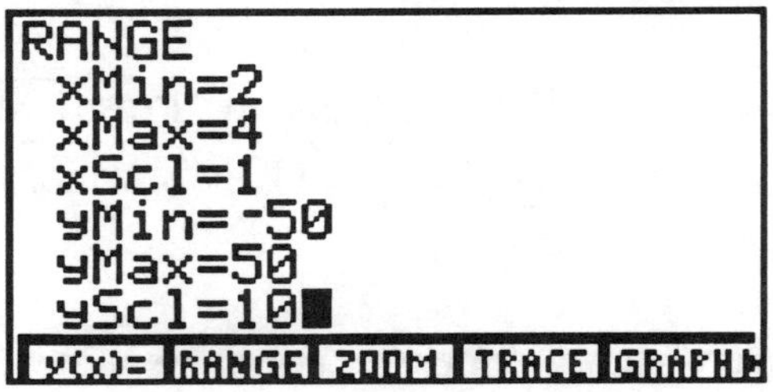

```
RANGE
 xMin=2
 xMax=4
 xScl=1
 yMin=-50
 yMax=50
 yScl=10
 y(x)= RANGE ZOOM TRACE GRAPH▶
```

This time portions of both branches of the graph were visible; however, the TI-85 did not draw the vertical asymptote.

Holes

Example: Graph $h(x) = \dfrac{x^2 - 4}{x - 2}$ in the two viewing windows specified below.

Window 1: Graph $f(x)$ in the standard viewing window.

Notice that $h(x)$ does not have a vertical asymptote! Instead, there is a hole in the graph at $x = 2$ that can't be seen in the present viewing screen.

Window 2: Graph $h(x)$ in the viewing window shown below.

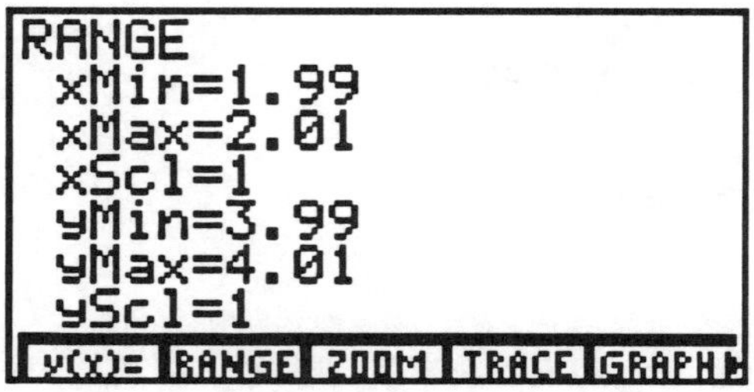

```
RANGE
 xMin=1.99
 xMax=2.01
 xScl=1
 yMin=3.99
 yMax=4.01
 yScl=1
 y(x)= RANGE ZOOM TRACE GRAPH▶
```

If you look carefully, you can now see the hole in the graph at (2, 4). (One pixel will be missing from the line.) The screen that follows shows what happens when you try to

110

get the coordinates for this hole using TRACE. To see why your calculator gives no value for y, try evaluating $h(2)$ yourself, by hand. What problem do you encounter?

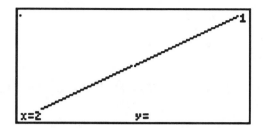

Project 4.5: Long-Term Behavior of Rational Functions

The Zoom Out feature on your calculator can be very useful for this project. See *Zooming Out*, page 107, for details.

When the Degree of the Denominator Is Greater Than or Equal To the Degree of the Numerator

Example: Explore the long-term behavior of the rational function $r(x) = \dfrac{6x+5}{2x-3}$.

Plan of action: Since the denominator of this function has the same degree as its numerator, we'll examine its long-term behavior by first graphing $r(x)$ in a viewing window that shows its key local features and then by zooming out in the x-direction only.

- Start by graphing $r(x)$ in the standard viewing window.

- Set the zoom factors: In the GRAPH/ZOOM menu, press MORE twice and then F1 for ZFACT. Set xFact to 4.

- Press F2 for RANGE and set xSscl to 0. (This will remove the tick marks from the x-axis.)

- Press F3 MORE MORE F2 for ZOOMX.

- Press ENTER to widen the width of the x-interval by a factor of 4. Then press ENTER a second and then third time to widen the interval by a factor of 16 and 64, respectively.

TI-85 Guide

- Press EXIT F4 for TRACE and use the right → and left ← arrow keys to move the cursor along the graph. (As you trace along the graph, the values of the y-coordinates should be approximately 3 except for points on the graph near the y-axis.)

When the Degree of the Denominator Is Less Than the Degree of the Numerator

Example: Explore the long-term behavior of the rational function $q(x) = \dfrac{4x^2 + x - 6}{x + 4}$.

Plan of action: First, we'll graph $q(x)$ in a viewing window that shows some of its local features. Then, since the degree of the denominator is less than that of the numerator, we'll examine the long-term behavior of $q(x)$ by zooming out in both the x- and y-directions.

- Graph $q(x)$ in the standard viewing window.

- Set the zoom factors: In the GRAPH/ZOOM menu, press MORE twice and then F1 for ZFACT. Set both xFact and yFact to 4.

- Press F2 for RANGE and set xSscl and yScl to 0. (This will remove the tick marks from the x- and y-axes.)

- Press F3 F3 for ZOUT. Then press ENTER to zoom out.

- Press ENTER two more times. (Wait until your calculator is finished with each zoom out before hitting ENTER again.) Your graph should resemble the line with positive slope shown below.

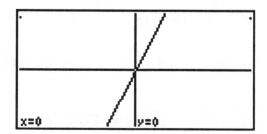

112

Chapter 5

General Information

Graphing Exponential Functions

In an exponential function, the independent variable is part of an exponent. You'll need to use the $\boxed{\wedge}$ key to enter the exponent. Also, because exponential functions increase or decrease very quickly in certain regions of their domain, you may have to experiment in order to find a viewing window that captures the function's important graphical features.

Example: Graph $y = 4^x$ in the standard viewing window.

To enter 4^x, press $\boxed{4}\,\boxed{\wedge}\,\boxed{F1}$.

Notice that, to the left of the y-axes, the graph appears to merge with the line $y = 0$ and to the right of the y-axes, the graph becomes so steep that it appears vertical.

Let's change the viewing window so that the graph will fill more of the viewing screen. Adjust the RANGE settings to match the ones below. (After graphing $y = 4^x$ in this viewing window, press $\boxed{\text{CLEAR}}$ to remove the menu bar from the bottom of the screen.)

```
RANGE
 xMin=-2
 xMax=2
 xScl=1
 yMin=0
 yMax=10
 yScl=1
 y(x)= RANGE ZOOM TRACE GRAPH
```

An exponential function has a positive number as its base (the base is the number that gets raised to the exponent). The next example should help you understand why we don't deal with negative bases.

Example: Graph $y = (-4)^x$ in the standard viewing window.

To enter $(-4)^x$, press $\boxed{(}\,\boxed{(\text{-})}\,\boxed{4}\,\boxed{)}\,\boxed{\wedge}\,\boxed{F1}$.

TI-85 Guide

Your calculator worked for a fairly long time before displaying a blank graph. To see why it had so much trouble, try evaluating the function for $x = .5, 1, 1.25, 1.5,$ and 2 using the TI-85's list feature as follows.

- Press $\boxed{\text{2nd}}$ $\boxed{\text{[QUIT]}}$ to return to the Home screen. Then press $\boxed{\text{CLEAR}}$.

- Press $\boxed{(}$ $\boxed{(-)}$ $\boxed{4}$ $\boxed{)}$ $\boxed{\wedge}$ $\boxed{\text{2nd}}$ $\boxed{\text{[LIST]}}$ $\boxed{\text{F1}}$ for {.

- Enter the x-values separated by commas and then press $\boxed{\text{F2}}$ for }.

- Press $\boxed{\text{ENTER}}$. Your screen should look like this:

```
(-4)^{.5,1,1.25,1.5,2
}
{(0,2) (-4,0) (-4,-4...
```
```
{  }  NAMES  EDIT  OPS
```

The final result is a list of values for $(-4)^{0.5}$, $(-4)^1$, $(-4)^{1.25}$, $(-4)^{1.5}$, and $(-4)^2$. Press the right arrow key $\boxed{\rightarrow}$ to observe the values that do not fit on the screen. Here's how to interpret the output:

Output for $(-4)^{0.5}$:
Recall that in radical notation $(-4)^{0.5}$ is equivalent to $\sqrt{-4}$. This is not a real number since there is no real number whose square is -4. However, you can re-express $\sqrt{-4}$ as $\sqrt{4} \cdot \sqrt{-1}$, $2\sqrt{-1}$, or $2i$. (The imaginary number i represents $\sqrt{-1}$.) Your TI-85 evaluated $(-4)^{0.5}$ as $(0, 2)$. This notation stands for the number $0 + 2\sqrt{-1}$.

Output for $(-4)^1$:
Your TI-85 evaluated $(-4)^1$ as $(-4, 0)$ which represents the real number $-4 + 0 \cdot \sqrt{-1}$ or -4.

Output for $(-4)^{1.25}$:
Your TI-85 evaluated $(-4)^{1.25}$ as $(-4, -4)$ which represents the complex number $-4 - 4 \cdot \sqrt{-1}$.

Notice that only $(-4)^1$ and $(-4)^2$ are real numbers.

Chapter 5

There are two bases for exponential functions, 10 and e, that are so common they have their own function keys on the calculator, $[10^x]$ and $[e^x]$.

Example: Graph $f(x) = e^x$ and $g(x) = 10^x$.

- First, erase any previously stored functions.

- In the GRAPH/$y(x) =$ menu, enter $f(x)$ as $y1$: press [2nd] [e^x] (same key as [LN]) [F1]. (Notice that you do <u>not</u> use the exponent key, [∧].)

- Enter $g(x)$ as $y2$: press [2nd] [10^x] (same key as [LOG]) [F1].

- Adjust your window settings to match the ones below.

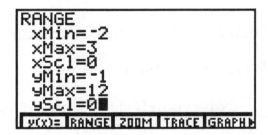

- Press [F5] to view the graphs.

Graphing Logarithmic Functions

The logarithmic functions with base e and base 10 have their own function keys [LN] and [LOG] on the calculator. Logarithmic functions of other bases can be graphed by dividing these functions by the appropriate scaling factor. (Refer to the Algebra Appendix, A.4(ii).)

Example: Graph $F(x) = \ln(x)$ and $G(x) = \log(x)$.

- First, erase any previously stored functions.

- In the GRAPH/$y(x) =$ menu, enter $F(x)$ as $y1$: press [LN] [F1].

- Enter $G(x)$ as $y2$: press [LOG] [F1].

TI-85 Guide

- Adjust your window settings to match the ones below.

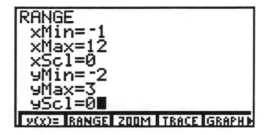

- Press $\boxed{F5}$ to view the graphs.

Lab 5A: AIDS

Graphing an Exponential Function Expressed in Base-e Form

Example: Graph $y = 3e^{.5x}$ in the standard viewing window.

- Press $\boxed{\text{GRAPH}}$ $\boxed{F1}$ for the GRAPH/$y(x) =$ menu. Clear any previously stored functions.

- To enter the function, press $\boxed{3}$ $\boxed{\text{2nd}}$ $\boxed{[e^x]}$ $\boxed{.}$ $\boxed{5}$ $\boxed{F1}$.

- To graph the function in the standard viewing window, press $\boxed{\text{EXIT}}$ $\boxed{F3}$ $\boxed{F4}$ for ZSTD.

Unlike many computer programs such as MAPLE, the TI-85 is not fussy about whether or not you enclose the exponent, $.5x$, in parenthesis. However, if you enter the exponent as $.5*x$, using the multiplication key, then you must enclose the exponent in parenthesis. Here is an example of what you cannot do:

- In the GRAPH/$y(x) =$ menu, enter the following function: press $\boxed{\text{2nd}}$ $\boxed{[e^x]}$ $\boxed{.}$ $\boxed{5}$ $\boxed{\times}$ $\boxed{F1}$.

- Press $\boxed{F5}$ to view the graph.

Although you may have expected an exponential function, your graph represents the linear function $y = (e^{.5})x$.

116

Warning! You must take care when interpreting the calculator-produced graphs of exponential functions. In places the graph is so steep that you might think that the function has a vertical asymptote. In other places the function is so close to zero that the graph in your viewing window merges with the x-axis, even though the function is never zero.

Lab 5B: Radioactive Decay

Graphing a Function Representing Exponential Decay

Example: Graph $y = e^{-\frac{x}{2}}$ in the standard viewing window.

There are two things that you must remember when entering this function. Use the $\boxed{(-)}$ key for the opposite of $\frac{x}{2}$ and enclose the exponent, $-\frac{x}{2}$, in parentheses. Your graph should match the one below.

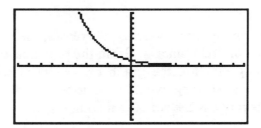

Lab 5C: Earthquakes

Tracing Beyond the Viewing Window

As with exponential functions, it is difficult to get a comprehensive picture of logarithmic functions from a single viewing window. For x-values near zero, the graph of a logarithmic function is so steep that it will appear to merge with the y-axis. As x-values become large, the graph becomes so flat that you might think that the function has a horizontal asymptote.

Example: Trace the x- and y-values on the graph of $y = \ln(x)$ beyond the viewing window.

- Clear any previously stored functions.

TI-85 Guide

- Enter the function ln(x) as $y1$: press [LN] [F1].

- Change the window settings to match the ones that follow and then press [F5].

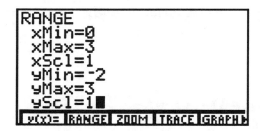

- Press [F4] for TRACE, and then press down the right arrow key [→]. As you continue to hold down the right arrow key [→], the window settings will automatically change to let you trace along a portion of the graph that lies outside the viewing window. Watch how slowly the y-values increase as the x-values increase. (Release the right arrow key from time to time, and observe the values of the x- and y-coordinates.)

Remember that $\ln(x)$ and e^x are inverses. Therefore, the range of $\ln(x)$ is the same as the domain of e^x, the set of real numbers. If, in the previous example, you press down on the right arrow key long enough, the graph of $y = \ln(x)$ will look constant in the viewing window. In such a viewing window, it looks as if ln (x) has a horizontal asymptote when, in fact, its range is the set of real numbers.

Chapter 6

General Information

Graphing Trigonometric Functions

Three of the six basic trigonometric functions are built-in functions on the TI-85: sine $\boxed{\text{SIN}}$, cosine $\boxed{\text{COS}}$, and tangent $\boxed{\text{TAN}}$. Before graphing any of these functions, you should first check that your calculator is set in radian mode: press $\boxed{\text{2nd}}$ $\boxed{\text{[MODE]}}$. If Radian is not already highlighted, move the cursor over Radian and press $\boxed{\text{ENTER}}$.

The standard window is generally not the best window to use when graphing trigonometric functions. Your calculator's trigonometric viewing window (ZTRIG) frequently provides a better scale to start with.

Example: Graph $y = \sin(x)$ in the trigonometric viewing window.

- Clear any previously stored function. Then in the GRAPH/$y(x) =$ menu, enter the function $\sin(x)$ as $y1$: press $\boxed{\text{SIN}}$ $\boxed{\text{F1}}$.

- To set the axes for the trig functions, press $\boxed{\text{EXIT}}$ $\boxed{\text{F3}}$ $\boxed{\text{MORE}}$ $\boxed{\text{F3}}$ for ZTRIG. Your graph should match the one that follows.

- Press $\boxed{\text{F2}}$ to observe the settings for the trig viewing window.

Problems Inherent in the Technology
(Don't Believe Everything That You See!)

Your viewing screen consists of a grid of pixels. (If you darken the contrast to its maximum setting, you may be able to see the grid.) When a pixel is *on*, it shows up as a

TI-85 Guide

dark square dot on the screen. Graphs are formed by turning on a series of pixels. This method of producing graphs can sometimes produce misleading images.

Let's look at what happens when we graph $y = \sin(x)$ over increasingly wide x-intervals.

- Start with a graph of sin(x) in the trigonometric viewing window. (Refer to the previous example.)

- Turn off the tick marks for the x-axis: enter the GRAPH/RANGE menu and change xScl to 0.

- Change the Zoom factors: press $\boxed{\text{F3}}$ to access the GRAPH/ZOOM menu, and then press $\boxed{\text{MORE}}$ $\boxed{\text{MORE}}$ $\boxed{\text{F1}}$ for ZFACT. Change the setting for XFact to 10.

- Press $\boxed{\text{F3}}$ $\boxed{\text{MORE}}$ $\boxed{\text{MORE}}$ $\boxed{\text{F2}}$ for ZOOMX.

Get ready for some fun. When sin(x) is graphed in the trig viewing window, you can observe two complete cycles (from -2π to 2π) of the wave. (In fact, between two and three complete cycles appear in the entire viewing window.) Each time you increase the width of the x-interval by a factor of ten, you should see ten times as many cycles of the sine wave.

- Press $\boxed{\text{ENTER}}$. Your graph should match the one below. Count the number of complete cycles of the sine wave that appear in this viewing window. (Your answer should be somewhere between 20 and 30.)

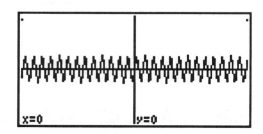

In this case, the calculator shows the correct number of cycles. However, because graphs on the calculator consist of a discrete set of highlighted pixels, your calculator is unable to produce the <u>smooth</u> wavy curve that is characteristic of sine waves. If you press enter again, you should expect ten times as many cycles as in the picture above, that is, between 200 and 300 cycles of sin(x).

- Now press $\boxed{\text{ENTER}}$ again to see what actually appears.

Chapter 6

This time the graph looks smooth with far <u>fewer</u> cycles than in the picture above! Here's what's going on. In producing this picture, your calculator does not have enough pixels to capture all the oscillations that are part of the actual graph. The small subset of points from the actual graph that the calculator chooses to represent with darkened pixels present a very misleading picture of the features of the actual graph!

Lab 6: Daylight and SAD

Modifying Basic Trigonometric Functions

Example: Graph $g(x) = \cos\left(x + \frac{\pi}{2}\right)$ in the trig viewing window.

To enter $\cos\left(x + \frac{\pi}{2}\right)$, press [COS] [(] [F1] [+] [2nd] [π] (same key as [^]) [÷] [2] [)].

The trigonometric viewing window gives a good picture of g. However, it is not the best window for viewing all trigonometric functions. Let's look at another example and use our understanding of how the constants in a trigonometric function affect its graph to help select a good viewing window.

Example: Graph $y = 5\cos(10x)$.

- First, look at the function in the trig viewing window.

The trig viewing window is not very good for displaying the key features of the function, so, we'll need to adjust the RANGE settings. Let's think about the effect that the constants 5 and 10 have on the basic cosine function. Recall from Projects 3.1 and 3.2 that multiplying a function by 5 stretches the basic graph vertically by a factor of 5 and that multiplying the input by 10 compresses the basic graph horizontally by a factor of 10. Thus, you will get a good viewing window for this function if you adjust the RANGE settings as follows: divide the x settings by 10 and multiply the y settings by 5. Try it!

- Press [F2] for RANGE. Change the x-settings to one-tenth of their present values and the y-settings to five times their present values.

- Press [F5] to observe the graph.

- Now press [F4] for TRACE. Use the right and left arrow keys to trace along the curve. What is the amplitude of this function?

121

Project 4: Don't *Lean* On Me

Away Dull Trig Tables!

With the TI-85 you can solve problems in right-triangle trigonometry without using trig tables. To compute the sine, cosine, or tangent of an angle measured in degrees, first change the Radian/Degree mode setting to Degree: press [2nd] [MODE] and highlight Degree by moving the cursor over Degree and pressing [ENTER].

Example: Compute sin 30° and tan 25°.

- Press [SIN] [3] [0] [ENTER]. (If you do not get .5 for the answer, go back and check that you have changed your calculator to degree mode.)

- Check that tan (25°) ≈ .466.

Example: Compute $\cos^{-1}(.5)$ and $\tan^{-1}(2.5)$.

- Press [2nd] [COS^{-1}] [.] [5] [ENTER]. Did you get approximately 60? Remember this is 60° if you're in degree mode.

- Check that $\tan^{-1}(2.5)$ ≈ 68.199°.

Finally, here's a way to compute the sine, cosine, or tangent of an angle given in degrees without changing the mode setting. First, return your calculator to its default mode setting, Radian: press [2nd] [MODE], highlight Radian, and press [ENTER]. Now, let's compute cos (60°) and sin (35°) without changing the mode setting. (This is a good idea because for most of your work in precalculus, you want radian mode.)

- Press [COS] [6] [0] [2nd] [MATH] [F3], for angle, [F1], to insert the degree symbol, [ENTER]. You should get .5 for your answer.

- Now use this method for sin (35°). Did you get approximately .574?

Chapter 7

General Information

Graphing Parametric Equations

For graphing parametric equations you'll need to change your calculator from function (Func) mode to parametric (Param) mode. Here's how: press $\boxed{\text{2nd}}$ $\boxed{\text{[MODE]}}$, then use the arrow keys to select **Param** and press $\boxed{\text{ENTER}}$.

Now let's see the changes in the GRAPH and GRAPH/RANGE menus.

- Press $\boxed{\text{GRAPH}}$.

- Press $\boxed{\text{F1}}$ to access the GRAPH/E(t) = menu. (Remember that in function mode, pressing $\boxed{\text{F1}}$ at this stage got you into the GRAPH/$y(x)$ = menu.) Notice that the input variable associated with the $\boxed{\text{F1}}$ key is t (instead of x).

- Press $\boxed{\text{EXIT}}$ to return to the GRAPH menu. Then press $\boxed{\text{F3}}$ $\boxed{\text{F4}}$ to set up the standard viewing window for parametric equations.

- Now press $\boxed{\text{F2}}$ to observe the RANGE settings. Use the up $\boxed{\uparrow}$ and down $\boxed{\downarrow}$ arrow keys to scroll through the entries in this menu. Notice that the settings associated with x and y are the same as they are in function mode. However, when your calculator is in parametric mode you must also specify bounds and increments for the parameter t: tMin, tMax, and tStep.

For example, suppose we want to graph the set of parametric equations

$$x(t) = 2t + 1, \qquad y(t) = -3t + 5$$

in the standard viewing window.

- Press $\boxed{\text{F1}}$ to enter the GRAPH/E(t) = menu. (Erase any previously stored parametric equations by positioning the cursor on the line containing the function and pressing $\boxed{\text{CLEAR}}$.)

- Enter the equation for x opposite $xt1$: press $\boxed{2}$ $\boxed{\text{F1}}$ $\boxed{+}$ $\boxed{1}$ $\boxed{\text{ENTER}}$.

TI-85 Guide

- Enter the equation for y opposite $yt1$: press $\boxed{(-)}$ $\boxed{3}$ $\boxed{F1}$ $\boxed{+}$ $\boxed{5}$.

- Press $\boxed{\text{EXIT}}$ $\boxed{F5}$ to graph the function. You should see a graph similar to the one that follows.

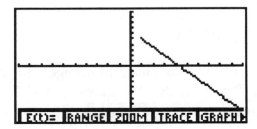

- Press $\boxed{F4}$ for TRACE. The cursor will mark the location when $t = 0$. The value of t and the coordinates of the point will appear at the bottom of your screen. Press the right arrow key $\boxed{\rightarrow}$ and the cursor will jump to the location associated with $t \approx .13$ (t increases by one tStep). Press $\boxed{\rightarrow}$ repeatedly and watch the cursor move along the line.

Next, let's graph the position of a dot as it moves along the path

$$x(t) = 2t + 1, y(t) = -3t + 5$$

at .5 second increments from time $t = 0$ seconds to $t = 3$ seconds.

For this example, we assume that your calculator is in parametric mode and that you have already entered this set of parametric equations in your calculator from the previous example.

- In the GRAPH menu, press $\boxed{\text{MORE}}$ $\boxed{F3}$ for FORMT (format). Move the cursor to DrawDot and press $\boxed{\text{ENTER}}$.

- Press $\boxed{F2}$ for RANGE. Adjust the parameter settings for t to match those on the screen that follows.

Chapter 7

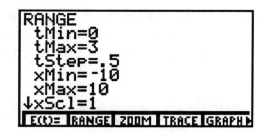

- Press $\boxed{F4}$ for TRACE. Then press $\boxed{\rightarrow}$ and watch the dot move from one position to the next (on a screen similar to the one below) in .5 second time increments.

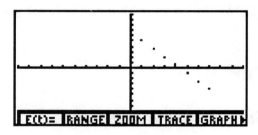

Lab 7A: Bordeaux

Graphing a Function With More Than One Input Variable

You can represent functions with more than one input variable graphically by replacing one (or more) of the independent variables with a list of values. (If you don't remember how to handle lists, refer to *Graphing a Family of Functions* on page 100.)

If you changed your calculator from the default settings, reset your calculator before starting this section:

- Press $\boxed{2nd}$ $\boxed{MEM}$ $\boxed{F3}$ for RESET.

- Now press $\boxed{F3}$ for DFLTS, and $\boxed{F4}$ for YES. (You may have to adjust your contrast. See page 1 if you have forgotten how.)

Example: Examine the behavior of $F(w, x) = 3w - 2x$ by holding w constant, first at -2, then at 0, and then at 2, while x varies.

TI-85 Guide

The three graphs, taken together, show how F varies with x for three different values of w. By forming a list $\{-2, 0, 2\}$ of the constant values for w, you can produce the graphs $y = F(-2, x)$, $y = F(0, x)$ and $y = F(2, x)$ using a single functional expression. Here's how:

- Press [GRAPH] [F1]. (Erase any previously stored functions.)

- Enter the function using the TI-85's list capabilities: press [3] [2nd] [[LIST]] [F1] for {, [(-)] [2] [,] [0] [,] [2] [F2] for }, [EXIT] (to return to the GRAPH/$y(x) =$ menu) [–] [2] [F1]. Notice that the input variable w has been replaced by the list of values $\{-2, 0, 2\}$.

- Press [EXIT] [F5] and watch as three parallel lines are graphed one by one.

Lab 7B: Bézier Curves

Combining Two Sets of Parametric Equations

In Lab 7B you are asked to form a new set of parametric equations from a combination of two other sets of parametric equations.

For example, suppose that you have two sets of parametric equations,

$S_1:$ $\quad x_1 = 2t + 1$
$\quad\quad\;\; y_1 = -3t + 5$

$S_2:$ $\quad x_2 = t - 5$
$\quad\quad\;\; y_2 = 4t - 3,$

and that you want to graph $(1 - t)S_1 + t\, S_2$, a combination of these equations, over the interval $0 \leq t \leq 1$.

Step 1: Change your calculator to parametric mode: press [2nd] [[MODE]], select **Param** and press [ENTER]. Check that your calculator is in DrawLine format: press [GRAPH] [MORE] [F3] select **DrawLine** and press [ENTER].

Step 2: Enter the two sets of parametric equations, S_1 and S_2.

- Erase any previously stored functions: press [2nd] [[MEM]] [F2] for DELETE, [MORE] [F3] for EQU. Using the down arrow key [↓], position the black

126

Chapter 7

marker opposite the first stored function and press ENTER. (This process will not remove RegEq.) Continue pressing ENTER until all functions are erased. Then press 2nd [QUIT].

- Press GRAPH F1 to access the GRAPH/E(t) = menu.

- Enter S_1's equations as $xt1$ and $yt1$ and then press ENTER. (Refer to *Graphing Parametric Equations*, page 123, if you have trouble.)

- Enter S_2's equations as $xt2$ and $yt2$ and press ENTER.

Step 3: Enter the x- and y- equations for the combination, $(1-t)S_1 + t\,S_2$:

$$x_3 = (1-t)x_1 + tx_2$$
$$y_3 = (1-t)y_1 + ty_2$$

- The cursor should be opposite $xt3$. Enter the x-equation for the combination:
  Press (1 − F1)
  Press F2 for xt and then 1 to get $xt1$.
  Press + F1
  Press F2 for xt and then 2 to get $xt2$.

- Enter the y-equation for the combination: move the cursor opposite $yt3$:
  Press (1 − F1)
  Press F3 for yt and then 1 to get $yt1$.
  Press + F1
  Press F3 for yt and then 2 to get $yt2$.

Step 4: Graph the combination, $(1-t)S_1 + t\,S_2$.

- Unselect (turn off) parametric equations $xt1$, $yt1$, $xt2$, and $yt2$. Move the cursor opposite $xt1$. Press F5. Notice that both $xt1$ and $yt1$ have been unselected. (The highlighting over their equals signs has been removed.) Next, move the cursor opposite $xt2$ and press F5.

- Press EXIT to return to the GRAPH menu. Press MORE F3 and check that DrawLine is highlighted.

TI-85 Guide

- Press F2 for RANGE. Set tMin $= 0$, tMax $= 1$, and tStep $= .1$. Adjust the remainder of the settings to match the ones below.

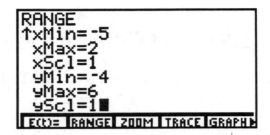

- Press F5 for GRAPH. Your graph should match this one:

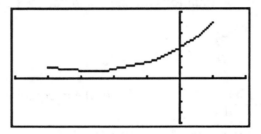

Project 7.3: What Goes Around Comes Around

Using Square Scaling in Parametric Mode

You may need to adjust the following settings:

- In the MODE menu (press 2nd MODE), check that Radian and Param are all highlighted.

- In the GRAPH/FORMT menu (press GRAPH MORE F3), check that SeqG and DrawDot are highlighted.

In addition to thinking about the x- and y-settings for the viewing window, you will need to adjust the RANGE settings for the parameter t. Use the following t-settings for all questions in this project except question 8.

- In the GRAPH menu, press F2 for RANGE.

- Set tMin to 0, tMax to 6.3, and tStep to .1.

Chapter 7

You will want to use square scaling for all your viewing windows.

- First, select a viewing window that shows the basic details of your graph.

- Then, in the GRAPH menu, change to square scaling by pressing [F3] [MORE] [F2] for ZSQR.

Example: Graph the set of parametric equations $x = 3\cos(t)$, $y = 3\sin(t) + 2$.

- Adjust the MODE and GRAPH/FORMT settings as indicated on the previous page.

- Clear any previously stored functions: press [2nd] [[MEM]] [F2] for Delete, and then press [MORE] [F3] for EQU. Move the marker opposite the first function and press [ENTER] repeatedly until all functions are erased.

- Press [GRAPH] [F1] and enter the set of parametric functions.

- In the GRAPH menu press [F2] for RANGE. Set tMin to 0, tMax to 6.3, and tStep to .1. Adjust the remaining settings to match the those on the screen below.

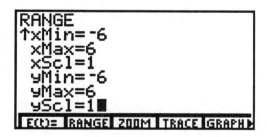

- Press [F5] to view the graph. Your graph should look egg-shaped. Next, press [F3] [MORE] [F2] for ZSQR and observe the graph in a window with square scaling. Your graph should resemble the one below.

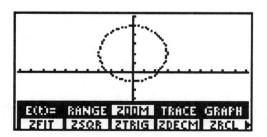